Heavy and Extra-heavy Oil Upgrading Technologies

Heavy and Extra-heavy Oil Upgrading Technologies

James G. Speight, PhD, DSc
CD&W Inc., Laramie, Wyoming, USA

ELSEVIER

AMSTERDAM • BOSTON • HEIDELBERG • LONDON
NEW YORK • OXFORD • PARIS • SAN DIEGO
SAN FRANCISCO • SINGAPORE • SYDNEY • TOKYO
Gulf Professional Publishing is an imprint of Elsevier

Gulf Professional Publishing is an imprint of Elsevier
The Boulevard, Langford Lane, Kidlington, Oxford, OX5 1GB, UK
25 Wyman Street, Waltham, MA 02451, USA

First published 2013

Notices
Knowledge and best practice in this field are constantly changing. As new research and experience broaden our understanding, changes in research methods, professional practices, or medical treatment may become necessary.

Practitioners and researchers must always rely on their own experience and knowledge in evaluating and using any information, methods, compounds, or experiments described herein. In using such information or methods they should be mindful of their own safety and the safety of others, including parties for whom they have a professional responsibility.

To the fullest extent of the law, neither the Publisher nor the authors, contributors, or editors, assume any liability for any injury and/or damage to persons or property as a matter of products liability, negligence or otherwise, or from any use or operation of any methods, products, instructions, or ideas contained in the material herein.

British Library Cataloguing-in-Publication Data
A catalogue record for this book is available from the British Library

Library of Congress Cataloging-in-Publication Data
A catalog record for this book is available from the Library of Congress

ISBN: 978-0-12-404570-5

For information on all Gulf Professional Publishing publications
visit our website at **store.elsevier.com**

This book has been manufactured using Print On Demand technology. Each copy is produced to order and is limited to black ink. The online version of this book will show color figures where appropriate.

Working together
to grow libraries in
developing countries

www.elsevier.com • www.bookaid.org

CONTENTS

The refining industry has been subjected to the impact of the four major forces that affect most industries and which have hastened the development of new petroleum refining processes, more specifically the changing quality of crude oil and geopolitics between different countries and the emergence of alternative feed supplies such as heavy oil, extra-heavy oil, and tar sand bitumen.

Difficult-to-refine feedstocks, such as heavy oil, extra-heavy oil, and tar sand bitumen, are characterized by low API gravity (high density) and high viscosity, high initial boiling point, high carbon residue, high nitrogen content, high sulfur content, and high metals content. In addition to these properties, the heavy feedstocks also have an increased molecular weight and reduced hydrogen content with a relatively low content of volatile saturated and aromatic constituents and a relatively high content of asphaltene and resin constituents, accompanied by a high heteroatom (nitrogen, oxygen, sulfur, and metals) content. Thus, such feedstocks are not typically subjected to distillation unless contained in the refinery feedstock as a blend with other crude oils.

The limitations of processing these heavy feedstocks depend to a large extent on the high molecular weight (low volatility) and heteroatom content and the tendency for coke formation and the deposition of metals and coke on the catalyst die. However, the essential step required of refineries is the upgrading of heavy. In fact, the increasing supply of heavy crude oils is a matter of serious concern for the petroleum industry. In order to satisfy the changing pattern of product demand, significant investments in refining conversion processes will be necessary to profitably utilize these heavy crude oils. The most efficient and economical solution to this problem will depend to a large extent on individual country and company situations.

This book presents viable options to the antiquated definitions of the heavy feedstocks (heavy oil, extra-heavy oil, tar sand bitumen) as well as an introduction to the various aspects of heavy feedstock refining in order for the reader to place each feedstock in the correct context of properties, behavior, and refining needs.

CHAPTER 1

Refining Heavy Oil and Extra-heavy Oil

1.1 HISTORY

The refining industry has been the subject of the four major forces that affect most industries and which have hastened the development of new petroleum refining processes: (i) the demand for products such as gasoline, diesel, fuel oil, and jet fuel; (ii) feedstock supply, specifically the changing quality of crude oil and geopolitics between different countries and the emergence of alternate feed supplies such as bitumen from tar sand, natural gas, and coal; (iii) environmental regulations that include more stringent regulations in relation to sulfur in gasoline and diesel; and (iv) technology development such as new catalysts and processes (Gary et al., 2007; Hsu and Robinson, 2006; Speight, 2011; Speight and Ozum, 2002).

In the early days of the twentieth century, refining processes were developed to extract kerosene for lamps. Any other products were considered to be unusable and were usually discarded. Thus, first refining processes were developed to purify, stabilize, and improve the quality of kerosene. However, the invention of the internal combustion engine led (at about the time of World War I) to a demand for gasoline for use in increasing quantities as a motor fuel for cars and trucks. This demand for the lower boiling products increased, particularly when the market for aviation fuel developed. Thereafter, refining methods had to be constantly adapted and improved to meet the quantity and quality requirements and needs of automobile and aircraft engines.

As the need for the lower boiling products developed, petroleum yielding the desired quantities of the lower boiling products became less available and refineries had to introduce conversion processes to produce greater quantities of lighter products from the higher boiling fractions. The means by which a refinery operates in terms of producing the relevant products, depends not only on the nature of the petroleum feedstock but also on its configuration (i.e., the number of types of the processes that are employed to produce the desired product slate) and the refinery configuration is, therefore, influenced by the specific demands of a market. Therefore, refineries need to be constantly adapted and upgraded to remain viable and responsive to ever changing patterns of crude supply and product market demands. As a result, refineries have been introducing increasingly complex and expensive processes to gain higher yields of lower boiling products from the heavy feedstocks.

The simplest refinery configuration is the *topping refinery*, which is designed to prepare feedstocks for petrochemical manufacture or for production of industrial fuels in remote oil-production areas. The topping refinery consists of tankage, a distillation unit, recovery facilities for gases and light hydrocarbons, and the necessary utility systems (steam, power, and water-treatment plants). Topping refineries produce large quantities of unfinished oils and are highly dependent on local markets, but the addition of hydrotreating and reforming units to this basic configuration results in a more flexible *hydroskimming refinery*, which can also produce desulfurized distillate fuels and high-octane gasoline. These refineries may produce up to half of their output as residual fuel oil, and they face increasing market loss as the demand for low-sulfur (even no-sulfur) and high-sulfur fuel oil increases.

The most versatile refinery configuration today is known as the *conversion refinery*. A conversion refinery incorporates all the basic units found in both the topping and the hydroskimming refineries, but it also features gas oil conversion plants, such as catalytic cracking and hydrocracking units, olefin conversion plants, such as alkylation or polymerization units, and, frequently, coking units for sharply reducing or eliminating the production of residual fuels. Modern conversion refineries may produce two-thirds of their output as unleaded gasoline, with the balance distributed between liquefied petroleum gas, jet fuel, diesel fuel, and a small quantity of coke.

This chapter presents an introduction to the definitions of the heavy feedstocks (heavy oil, extra-heavy oil, and tar sand bitumen) as well as to the various aspects of petroleum refining in order for the reader to place each feedstock in the correct context of properties, behavior, and refining needs (Speight, 2006a,b, 2011).

1.2 DEFINITION OF THE FEEDSTOCKS

Heavy oil reservoirs, extra-heavy oil reservoirs, and tar sand deposits are widely distributed throughout the world in a variety of countries with much of the known heavy oil and extra-heavy oil reservoirs occurring in the United States and Venezuela and the major known tar sand deposits occurring in Canada (Speight, 2007, 2009, 2013).

1.2.1 Heavy Oil

Heavy oil is a *type* of petroleum that is different from conventional petroleum insofar as it is much more difficult to recover from the subsurface reservoir (Speight, 2007, 2009). These materials have a much higher viscosity (and lower API gravity) than conventional petroleum and recovery of these petroleum types usually requires thermal stimulation of the reservoir.

Heavy oil is a petroleum-type resource that is characterized by high viscosities (i.e., resistance to flow) and high densities compared to conventional oil. Most heavy oil reservoirs originated as conventional oil that formed in deep formations but migrated to the surface region where they were degraded by bacteria and by weathering, and where the lightest hydrocarbons escaped. Heavy oil is deficient in hydrogen and has high carbon, sulfur, and heavy metal content. Hence, heavy oil requires additional processing (upgrading) to become a suitable refinery feedstock for a normal refinery.

However, heavy oil is more difficult to recover from the subsurface reservoir than conventional or light oil. A very general definition of heavy oils has been based on the API gravity or viscosity, and the definition is quite arbitrary although there have been attempts to rationalize the definition based upon viscosity, API gravity, and density.

The term *heavy oil* has also been arbitrarily (but incorrectly) used to describe both the heavy oils that require thermal stimulation of recovery from the reservoir and the bitumen in bituminous sand (tar sand) formations from which the heavy bituminous material is recovered by a mining operation.

1.2.2 Extra-heavy Oil

The term *extra-heavy oil* is a recently evolved term (related to viscosity) of little scientific meaning. While this type of oil may resemble tar sand bitumen and does not flow easily, extra-heavy oil is generally recognized as having mobility in the reservoir compared to tar sand bitumen, which is typically incapable of mobility (free flow) under reservoir conditions. For example, the tar sand bitumen located in

Alberta, Canada is not mobile in the deposit and requires extreme methods of recovery to recover the bitumen. On the other hand, much of the extra-heavy oil located in Orinoco, belt of Venezuela requires recovery methods that are less extreme because of the mobility of the material in the reservoir (Schenk et al., 2009; Total SA, 2007).

Whether the mobility of extra-heavy oil is due to a high reservoir temperature (i.e., higher than the pour point of the extra-heavy oil) or due to other factors is variable and subject to localized conditions in the reservoir.

1.2.3 Tar Sand Bitumen

Tar sand bitumen is another source of liquid fuels that is distinctly separate from conventional petroleum (Speight, 2005a,b, 2007, 2009; US Congress, 1976). Understanding the definition of tar sand bitumen is essential in placing definitions of the terms heavy oil and extra-heavy oil in the correct scientific and engineering perspective.

Tar sand, also called *oil sand* (in Canada), or the more geologically correct term *bituminous sand*, is commonly used to describe a sandstone reservoir that is impregnated with a heavy, viscous bituminous material. Tar sand is actually a mixture of sand, water, and bitumen but many of the tar sand deposits in countries other than Canada lack the water layer that is believed to facilitate the hot water recovery process. The heavy bituminous material has a high viscosity under reservoir conditions and cannot be retrieved through a well by conventional production techniques.

Geologically, the term *tar sand* is commonly used to describe a sandstone reservoir that is impregnated with bitumen, a naturally occurring material that is solid or near solid and is substantially immobile under reservoir conditions. The bitumen cannot be retrieved through a well by conventional production techniques, including currently used enhanced recovery techniques. In fact, tar sand is defined (FEA-76-4) in the United States Congress as:

> The several rock types that contain an extremely viscous hydrocarbon which is not recoverable in its natural state by conventional oil well production methods including currently used enhanced recovery techniques. The hydrocarbon-bearing rocks are variously known as bitumen-rocks oil, impregnated rocks, tar sands, and rock asphalt.

By inference, conventional petroleum and heavy oil are recoverable by well-production methods (i.e., primary and secondary recovery methods) (Speight, 2009, 2013) and by currently used enhanced oil recovery (EOR) methods (Speight, 2007, 2009, 2013).

However, the term *tar sand* is actually a misnomer; more correctly, the name *tar* is usually applied to the heavy product remaining after the destructive distillation of coal or other organic matter (Speight, 2007, 2008, 2011, 2013). Current recovery operations of bitumen in tar sand formations have been focused predominantly on a mining technique but thermal *in situ* processes are now showing success (Speight, 2009).

In addition to the above definitions, there are several tests that must be carried out to determine whether or not, in the first instance, a resource is a tar sand deposit (Speight, 2001). Most of all, a core taken from a tar sand deposit and the bitumen isolated therefrom are certainly not identifiable by the preliminary inspections (sight and touch) alone. In the United States, the final determinant is whether or not the material contained therein can be recovered by primary, secondary, or tertiary (enhanced) recovery methods (US Congress, 1976).

The precise chemical composition of any heavy feedstock is, despite the large volume of work performed in this area, largely speculative. In very general terms (and as observed from elemental analyses), bitumen is a complex mixture of (i) hydrocarbons, (ii) nitrogen compounds, (iii) oxygen compounds, (iv) sulfur compounds, and (v) metallic constituents. However, this general definition is not adequate to describe the composition of bitumen as it relates to behavior.

The properties of heavy oil, extra-heavy oil, and tar sand bitumen fall into a wide range although the properties are often comparable to those of atmospheric residual and vacuum residual and several inter-property relationships can be observed (Speight, 2007). However, the properties of each of these feedstocks have been presented elsewhere (Ancheyta and Speight, 2007; Speight, 2007, 2008, 2009, 2011, 2013), to which the reader is referred for further details.

1.3 REFINING HEAVY OIL AND EXTRA-HEAVY OIL

The demand for petroleum and petroleum products has shown a sharp growth in recent years (Hsu and Robinson, 2006; Speight, 2007, 2011);

this could well be the last century for petroleum refining, as we know it. The demand for transportation fuels and fuel oil is forecast to continue to show a steady growth in the future. The simplest means to cover the demand growth in low-boiling products is to increase the imports of light crude oils and low-boiling petroleum products, but these steps may be limited in the future.

Over the past three decades, crude oils available to refineries have generally decreased in API gravity. There is, nevertheless, a major focus in refineries on the ways in which heavy feedstocks might be converted into low-boiling high-value products (Khan and Patmore, 1997). Simultaneously, the changing crude oil properties are reflected in changes such as an increase in asphaltene constituents, and an increase in sulfur, metal, and nitrogen contents. Pretreatment processes for removing such constituents or at least negating their effect in thermal process would also play an important role.

Difficult-to-refine feedstocks, such as heavy oil, extra-heavy oil, and tar sand bitumen, are characterized by low API gravity (high density) and high viscosity, high initial boiling point, high-carbon residue, high-nitrogen content, high-sulfur content, and high-metals content (Speight, 2007, 2013). In addition to these properties, the heavy feedstocks also have an increased molecular weight and reduced hydrogen content (Figure 1.1) with a relatively low content of volatile saturated and aromatic constituents and a relatively high content of asphaltene and resin constituents that is accompanied by a high-heteroatom (nitrogen, oxygen, sulfur, and metals) content (Figure 1.2). Thus, such feedstocks are not typically subject to distillation unless contained in the refinery feedstock as a blend with other crude oils.

The limitations of processing these heavy feedstocks depend to a large extent on the tendency for coke formation and the deposition of metals and coke on the catalyst due to the higher molecular weight (low volatility) and heteroatom content. However, the essential step required of refineries is the upgrading of heavy feedstocks (Dickenson et al., 1997; McKetta, 1992). In fact, the increasing supply of heavy crude oils is a matter of serious concern for the petroleum industry. In order to satisfy the changing pattern of product demand, significant investments in refining conversion processes will be necessary to profitably utilize these heavy crude oils. The most efficient and economical

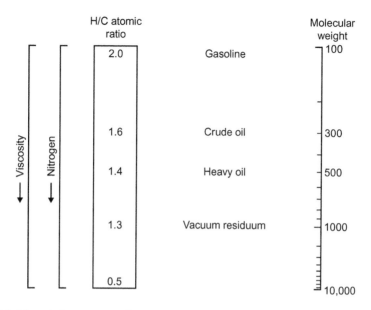

Figure 1.1 Relative hydrogen content (through the atomic H/C ratio) and molecular weight of refinery feedstocks—heavy oil, extra-heavy oil, and tar sand bitumen generally fall into the residuum range.

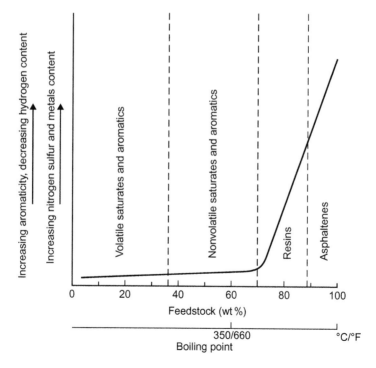

Figure 1.2 Relative distribution of heteroatoms in the various fractions.

solution to this problem will depend to a large extent on individual country and company situations.

Upgrading heavy feedstock began with the introduction of desulfurization processes (Ancheyta and Speight, 2007; Hsu and Robinson, 2006; Speight, 2000, 2007). In the early days, the goal was desulfurization but, in later years, the processes were adapted to a 10−30% partial conversion operation, as intended to achieve desulfurization and obtain low-boiling fractions simultaneously, by increasing severity in operating conditions. Refinery evolution has seen the introduction of a variety of cracking processes based on *thermal cracking, catalytic cracking,* and *hydroconversion.* Those processes are different from one another in cracking method, cracked product patterns, and product properties and will be employed in refineries according to their respective features. Thus, refining heavy feedstocks has become a major issue in modern refinery practice and several process configurations have evolved to accommodate the heavy feedstocks (Table 1.1) (Gary et al., 2007; Hsu and Robinson, 2006; Khan and Patmore, 1997; RAROP. 1991; Shih and Oballa, 1991; Speight, 2007).

Technologies for upgrading heavy feedstocks can be broadly divided into *carbon rejection* and *hydrogen addition* processes. *Carbon rejection* redistributes hydrogen among the various components, resulting in fractions with increased hydrogen/carbon (H/C) atomic ratios and fractions with decreased H/C atomic ratios. On the other hand,

Table 1.1 Selected Examples of Recent Processes Configurations for Refining Heavy Feedstocks	
Thermal processes	**Comment**
ASCOT process	Delayed coking with deep solvent deasphalting
CHERRY-P process	Feedstock-coal slurry
ET-II process	Feedstock mixed with high-boiling recycle oil
Catalytic cracking processes	
ART process	Efficient feedstock-catalyst contact
HOC process	Feedstock: desulfurized residuum
HOT process	Hydrogen produced in the unit by steam−hydrogen reaction
Solvent processes	
DEMEX process	Less selective solvent than propane
MDS process	Solvent deasphalting and/or desulfurization of feedstock
ROSE process	Deasphaltening of feedstock

hydrogen addition processes involve reaction of heavy crude oils with an external source of hydrogen and result in an overall increase in H/C ratio (Speight, 2007, 2011; Stanislaus and Cooper, 1994).

The criteria to select one of the routes as an upgrading option depends on several factors which must be analyzed in detail when it comes to consider a project of this nature. For example, the technology of *hydrogen addition* produces a high yield of products with a commercial value larger than that of the *carbon rejection* technology but requires a larger investment and more natural gas availability to produce the amounts of hydrogen and steam required for these processes.

New processes for the conversion of heavy oil feedstocks will probably be used in concert with visbreaking with some degree of hydroprocessing as a primary conversion step. Other processes may replace or augment the deasphalting units in many refineries. Depending on the properties, an option for heavy oil, like the early option for tar sand bitumen, is to subject the feedstock to either delayed coking or fluid coking as the *primary upgrading* step (Figure 1.3) with some prior distillation or topping (Speight, 2007, 2009, 2011). After primary upgrading, the product steams are hydrotreated and combined to form a *synthetic crude oil* that is shipped to a conventional refinery for further processing to liquid fuels.

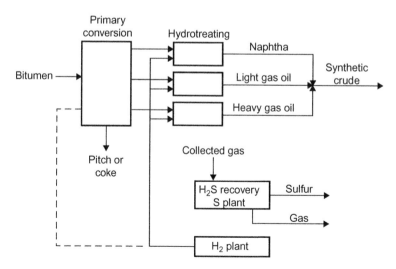

Figure 1.3 Example of a processing sequence for tar sand bitumen.

However, there is not one single heavy oil upgrading solution that will fit all refineries. Heavy feedstock properties, existing refinery configuration, and desired product slate all can have a significant effect on the final configuration. Furthermore, a proper evaluation is not a simple undertaking for an existing refinery. The evaluation starts with an accurate understanding of the nature of the feedstock along with corresponding conversion chemistry assesment. Once the options have been defined, development of the optimal configuration for refining the incoming feedstocks can be designed.

There is also recognition that *in situ* upgrading could be a very beneficial process for leaving the unwanted elements in the reservoir and increasing the API gravity. While this is important in the context of this text insofar as upgrading during recovery will reduce surface refining costs, it is described in more detail elsewhere and will not be repeated here. However, the salient facts of the concept are worthy of note.

There are two ways that are currently practiced in bringing heavy crude oil to market. The first method is to upgrade the material in the oil field and leave much of the materials behind as coke and then pipeline the upgraded material out as synthetic crude. In this method, the crude is fractionated and the residue is coked. The products of the coking operation, and in some cases some of the residue, are hydrotreated. The hydrotreated materials are recombined with the fractionated light materials to form synthetic crude that is then transported to market in a pipeline.

The second method is to affect partial upgrading *in situ* as part of the recovery process. Such an option to produce an acceptable pipeline material would be an ideal solution but has a number of limitations (Motaghi et al., 2010a,b,c; Speight, 2013). For example, the amount of heavy oil production could be limited by the recovery process and the upgraded products must be compatible with the original or partially changed heavy oil (Speight, 2009). If the products and the original (partially changed) heavy oil have limited compatibility, that would limit the amount of dilution and again could limit the effectiveness of the refining process (Speight, 2007, 2011).

However, the increased mobilization of heavy oil in the reservoir by partial upgrading is not a new idea and still has many hurdles to overcome before it can be considered close to commercial. The product will be less viscous than the heavy oil in place but some property changes such as high-olefin content from cracking are not necessarily positive. In summary, there are three main approaches for heating the reservoir: (i) steam distillation, (ii) mild thermal cracking—visbreaking, and (iii) partial combustion (Speight, 2009, 2013).

Nevertheless, there is (or will be) an obvious future need for partial upgrading during or immediately after recovery. On the other hand, hydrogen addition must be used during upgrading in order to stabilize the upgraded heavy oil—which could mean that the cost of partial upgrading is not much reduced as compared to full upgrading. Therefore, the only choice currently is no upgrading or full upgrading. Other goals could be to achieve breakthroughs in upgrading technologies—such as nonthermal coking methods that would use far less energy or gasification at 800°C (1470°F), which is far lower than current commercial temperatures. The technology where changes do occur involves combustion of the oil *in situ*. The concept of any combustion technology requires that the oil be partially combusted and that thermal decomposition occur to other parts of the oil. This is sufficient to cause irreversible chemical and physical changes to the oil to the extent that the product is markedly different to the oil in place. Recognition of this phenomenon is essential before combustion technologies are applied to oil recovery.

Although this improvement in properties may not appear to be too drastic, nevertheless it is usually sufficient to have major advantages for refinery operators. Any incremental increase in the units of H/C ratio can save useful quantities of costly hydrogen during upgrading. The same principles are also operative for reductions in the nitrogen, sulfur, and oxygen contents. This latter occurrence also improves catalyst life and activity as well as reduces the metals content.

In short, *in situ* recovery processes (although less efficient in terms of bitumen recovery relative to mining operations) may have the added benefit of *leaving* some of the more obnoxious constituents (from the processing objective) in the ground. Processes that offer the potential for partial upgrading during recovery are varied but usually follow a surface process. Not that this be construed as an easy task; there are many disadvantages that arise from attempting *in situ* upgrading.

Finally, there is not a single *in situ* recovery process that will be applicable to all reservoirs and no single recovery process will be able to access all the heavy oil in a given reservoir. To achieve maximum recovery, it will be necessary to apply a combination of different processes. For example, by use of a steam-based recovery process followed by an *in situ* combustion, then by *in situ* upgrading, and by bioconversion of the residual hydrocarbons. This type of sequential recovery will require careful planning to ensure that the optimum sequence and timing is applied.

Then, to achieve partial upgrading during recovery requires a further sequential operation before a transportable is produced. A multistep system is required to achieve the necessary aims of heavy oil recovery with partial upgrading. What this might be is currently unknown but there are possibilities.

Thus, a major decision at the time of recovery of heavy oil, extra-heavy oil, and tar sand bitumen is to acknowledge the practical (or impractical) aspects of upgrading during recovery, partial upgrading at the surface, or full upgrading in a conversion refinery. For the purposes of this text only, upgrading in a conversion refinery has been assumed as the means of upgrading one or all of the three heavy feedstocks.

REFERENCES

Ancheyta, J., Speight, J.G., 2007. Hydroprocessing Heavy Oils and Heavy Feedstocks. CRC Press, Taylor & Francis Group, Boca Raton, FL.

Dickenson, R.L., Biasca, F.E., Schulman, B.L., Johnson, H.E., 1997. Refiner options for converting and utilizing heavy fuel oil. Hydrocarbon Process. 76 (2), 57–62.

Gary, J.H., Handwerk, G.E., Kaiser, M.J., 2007. Petroleum Refining: Technology and Economics, fifth ed. CRC Press, Taylor & Francis Group, Boca Raton, FL.

Hsu, C.S., Robinson, P.R. (Eds.), 2006. Practical Advances in Petroleum Processing, vols. 1 and 2. Springer, New York, NY.

Khan, M.R., Patmore, D.J., 1997. Heavy oil upgrading processes. In: Speight, J.G. (Ed.), Petroleum Chemistry and Refining. Taylor & Francis, Washington, DC (Chapter 6).

McKetta, J.J. (Ed.), 1992. Petroleum Processing Handbook. Marcel Dekker, New York, NY.

Motaghi, M., Shree, K., Krishnamurthy, S., 2010a. Consider new methods for bottom of the barrel processing. Part 1: advanced methods use molecule management to upgrade heavy ends. Hydrocarbon Process. 88 (2), 35–38.

Motaghi, M., Shree, K., Krishnamurthy, S., 2010b. Consider new methods for bottom of the barrel processing. Part 2: new methods of molecule management dictate the best economics when upgrading residuum. Hydrocarbon Process. 88 (3), 55–58.

Motaghi, M., Saxena, P., Ravi, R., 2010c. Partial upgrading of heavy oil reserves. Pet. Technol. Q. Q4, 55–64.

RAROP, 1991. RAROP Heavy Oil Processing Handbook. T. Noguchi (Chairman). Research Association for Residual Oil Processing. Ministry of Trade and International Industry (MITI), Tokyo, Japan.

Schenk, C.J., Cook, T.S.A., Charpentier, R.R., Pollastro, R.M., Klett, T.R. Tennyson, M.E., et al., 2009. An Estimate of Recoverable Heavy Oil Resources of the Orinoco Oil Belt, Venezuela. Fact Sheet 2009—3028. United States Geological Survey, US Department of the Interior, Reston, Virginia. October.

Shih, S.S., Oballa, M.C. (Eds.), 1991. Tar Sand Upgrading Technology. American Institute for Chemical Engineers, New York, NY (Symposium Series No. 282).

Speight, J.G., 2000. The Desulfurization of Heavy Oils and Residua, second ed. Marcel Dekker, New York, NY.

Speight, J.G., 2001. Handbook of Petroleum Analysis. John Wiley & Sons Inc., New York.

Speight, J.G., 2005a. Chemistry and Physics of Natural Bitumen and Heavy Oil, in Coal, Oil Shale, Natural Bitumen, Heavy Oil and Peat, from Encyclopedia of Life Support Systems (EOLSS), Developed Under the Auspices of the UNESCO. EOLSS Publishers, Oxford, UK <http://www.eolss.net> (accessed 31.01.13).

Speight, J.G., 2005b. Upgrading and Refining of Natural Bitumen and Heavy Oil, in Coal, Oil Shale, Natural Bitumen, Heavy Oil and Peat, from Encyclopedia of Life Support Systems (EOLSS), Developed Under the Auspices of the UNESCO. EOLSS Publishers, Oxford, UK <http://www.eolss.net> (accessed 31.01.13).

Speight, J.G., 2007. The Chemistry and Technology of Petroleum, fourth ed. CRC Press, Taylor & Francis Group, Boca Raton, FL.

Speight, J.G., 2008. Synthetic Fuels Handbook: Properties, Processes, and Performance. McGraw-Hill, New York, NY.

Speight, J.G., 2009. Enhanced Recovery Methods for Heavy Oil and Tar Sands. Gulf Publishing Company, Houston, TX.

Speight, J.G., 2011. The Refinery of the Future. Gulf Professional Publishing, Elsevier, Oxford, UK.

Speight, J.G., 2013. Oil Sand Production Processes. Gulf Professional Publishing, Elsevier, Oxford, UK.

Speight, J.G., Ozum, B., 2002. Petroleum Refining Processes. Marcel Dekker, New York, NY.

Stanislaus, A., Cooper, B.H., 1994. Catal. Rev. — Sci. Eng. 36 (1), 75.

Total S.A., 2007. Extra Heavy Oils and Bitumen Reserves for the Future. TOTAL S.A. Exploration & Production, Pau, France.

US Congress, 1976. Public Law FEA-76-4. United States Congress, Library of Congress, Washington, DC.

CHAPTER 2

Thermal Cracking

2.1 INTRODUCTION

Thermal cracking was one of the first conversion processes used in the oil industry and has been employed since 1913, when different fuels and heavy hydrocarbons were heated under pressure in large drums until reaching their thermal fracture into lower molecular size products with a lower boiling point.

Thermal cracking processes offer attractive methods of conversion of heavy feedstocks because they enable low operating pressure, while involving high operating temperature, without requiring expensive catalysts. Currently, the most widely operated heavy feedstock conversion processes are visbreaking and delayed coking. And, these are still attractive processes for refineries from an economic point of view and will continue to be so well into the twenty-first century (Dickenson et al., 1997; Gary et al., 2007; Hsu and Robinson, 2006; Speight, 2007, 2011).

The majority of regular thermal cracking processes use temperatures of 455–540°C (850–1005°F) and pressures of 100–1000 psi. This approach is the oldest technology available for residue conversion, and the severity of thermal processing determines the conversion and the product characteristics. Thermal treatment of heavy feedstock ranges from mild treatment for reduction of viscosity to *ultrapyrolysis* (high-temperature cracking at very short residence time) for better conversion to overhead products (Hulet et al., 2005). A higher temperature requires a shorter time to achieve a given conversion but, in many cases, there is *a change in the chemistry of the reaction* so merely raising the temperature does not necessarily accomplish the same goals in terms of product slate and product yields.

Low pressures (<100 psi) and temperatures in excess of 500°C (930°F) tend to produce lower molecular weight hydrocarbons than those produced at higher pressures (400–1000 psi) and at temperatures below 500°C (930°F). The reaction time is also important; light feeds (gas oils) and recycle oils require longer reaction times than the readily cracked heavy residues. Mild cracking conditions (defined here as a low conversion per cycle) favor a high yield of gasoline components with decreased gas production and decreased coke production, but the gasoline quality is not high, whereas more severe conditions give increased gas production and increased coke production and reduced gasoline yield (but of higher quality). With limited conversion per cycle, the heavier residues must be recycled. However, the recycled oils become increasingly refractory upon repeated cracking, and if they are not required as a fuel oil stock they may be subjected to a coking operation to increase the gasoline yield or refined by means of a hydrogen process.

Although new thermal cracking units are now under development for heavy feedstocks (Speight, 2007, 2011), processes that can be regarded as having evolved from the original concept of thermal cracking are visbreaking and the various coking processes (Table 2.1). It is the purpose of this chapter to present these processes in the light of their use in modern refineries and the information that should be borne in mind when considering and deciding upon the potential utility of any process presented throughout this and the subsequent chapters.

The importance of solvents to mitigate coke formation has been recognized for many years, but their effects have often been ascribed to hydrogen-donor reactions rather than phase behavior. The separation of the phases depends on the solvent characteristics of the liquid. Addition of aromatic solvents will suppress phase separation (Speight, 2007), while paraffin-based solvents will enhance separation.

In summary, there is a need to improve conversion of heavy feedstocks and part of the future growth will be at or near recovery sites at heavy crude reservoirs and bitumen deposits in order to improve the quality to ease transportation and open markets for crudes of otherwise marginal value.

The purpose of this chapter is to present descriptions of the thermal processes that are available for the conversion of heavy feedstocks and to place them in the perspective of the future refinery.

2.2 THERMAL CRACKING PROCESSES

The majority of the thermal cracking processes are carried out using relatively simple reactors at relatively low pressure and at temperatures

Table 2.1 Comparison of Visbreaking with Delayed Coking and Fluid Coking

Thermal cracking
Purpose: To produce volatile products of low-volatile or nonvolatile feedstocks
Conversion is the prime purpose
Cracking with simultaneous removal of distillate (semicontinuous)
Batch cracking (noncontinuous)
High conversion
Process configuration: Various
Visbreaking
Purpose: To reduce viscosity of fuel oil to acceptable levels
Conversion is not a prime purpose
Mild (470−495°C; 880−920°F) heating at pressures of 50−200 psi
Reactions quenched before going to completion
Low conversion (10%) to products boiling less than 220°C (430°F)
Heated coil or drum (soaker)
Delayed coking
Purpose: To produce maximum yields of distillate products
Moderate (480−515°C; 900−960°F) heating at pressures of 90 psi
Reactions allowed to proceed to completion
Complete conversion of the feedstock
Soak drums (845−900°F) used in pairs (one on stream and one off stream being decoked)
Coked until drum solid
Coke removed hydraulically from off-stream drum
Coke yield: 20−40% by weight (dependent upon feedstock)
Yield of distillate boiling below 220°C (430°F): *ca.* 30% (but feedstock dependent)
Fluid coking
Purpose: To produce maximum yields of distillate products
Severe (480−565°C; 900−1050°F) heating at pressures of 10 psi
Reactions allowed to proceed to completion
Complete conversion of the feedstock
Oil contacts refractory coke
Bed fluidized with steam; heat dissipated throughout the fluid bed
Higher yields of light ends ($<C_5$) than delayed coking
Less coke-make than delayed coking (for one particular feedstock)

in the range 450−510°C (840−920°F)—where temperatures are not presented here given it is because they were not given by the reactor developer and this temperature range is to be assumed as the operative process temperature range.

2.2.1 Visbreaking

Visbreaking (i.e., viscosity breaking, viscosity reduction) was originally developed to produce specification-grade fuel oil from residua but is often considered to be a conversion process and is a relatively mild thermal cracking operation (Dominici and Sieli, 1997; Gary et al., 2007; Hsu and Robinson, 2006; Speight 2007, 2011; Speight and Ozum, 2002). By reducing the viscosity of the nonvolatile fraction, visbreaking produces distillate material that can either be used as a fuel (prior to coking the nonvolatile constituents) or as a solvent so that total visbroken product is more amenable to pipeline transportation or for use as a feedstock to a catalytic cracking process.

The visbreaking process (Speight and Ozum, 2002) uses the approach of mild thermal cracking as a relatively low cost and low-severity approach to improving the viscosity characteristics of the residue without attempting significant conversion to distillates.

Visbreaking conditions range from 455°C to 510°C (850–950°F) at a short residence time and from 50 to 300 psi at the heating coil outlet. A short residence time is required to avoid coke formation, and it is the short residence time that brings to visbreaking the concept of being a mild thermal reaction in contrast to, for example, the delayed coking process where residence times are much longer and the thermal reactions are allowed to proceed to completion. The visbreaking process uses a quench operation to terminate the thermal reactions. Additives are used on occasion when the feedstock is particularly prone to deposition of thermal coke, which forms on the furnace heating coils.

In a typical process configuration (Figure 2.1), a heavy feedstock is passed through a furnace where it is heated to a temperature of 480°C (895°F) under an outlet pressure of about 100 psi. The heating coils in the furnace are arranged to provide a soaking section of low-heat density, where the charge remains until the visbreaking reactions are completed. The cracked products are then passed into a flash-distillation chamber. The overhead material from this chamber is then fractionated to produce naphtha as an overhead product and light gas oil. The liquid products from the flash chamber are cooled with a gas oil flux and then sent to a vacuum fractionator. This yields a heavy gas oil distillate and a heavy feedstock of reduced viscosity (Table 2.2). A 5–10% conversion of the heavy feedstock to naphtha is usually

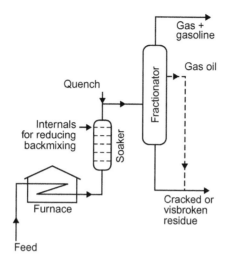

Figure 2.1 A soaker visbreaker.

	Arabian Light Vacuum	Arabian Light Vacuum	Iranian Light Vacuum	Athabasca
Table 2.2 Examples of Product Yields and Properties from Visbreaking Athabasca Tar Sand Bitumen and Feedstocks Having a Similar API Gravity				
Feedstock	Heavy feedstock	Heavy feedstock	Residue	Bitumen
API gravity	7.1	6.9	8.2	8.6
Carbon residue[a]	20.3		22.0	13.5
Sulfur, wt. %	4.0	4.0	3.5	4.8
Product yields[b], vol. %				
Naphtha (<425°F, <220°C)	6.0	8.1	4.8	7.0
Light gas oil (425−645°F, 220−340°C)	16.0	10.5	13.1	21.0
Heavy gas oil (645−1000°F, 340−540°C)		20.8	b	35.0
Heavy feedstock	76.0	60.5	79.9	34.0
API gravity	3.5	0.8	5.5	
Carbon residue[a]				
Sulfur, wt. %	4.7	4.6	3.8	

[a] Conradson.
[b] A blank product yield line indicates that the yield of the lower boiling product has been included in the yield of the higher boiling product.

sufficient to afford at least an approximate fivefold reduction in viscosity. Reduction in viscosity is also accompanied by a reduction in the pour point.

In the context of this book, an alternative option for the heavy feedstocks is to use lower furnace temperature and longer time achieved by installing a soaking drum between the furnace and the fractionator. The disadvantage of this approach is the need for careful monitoring of the process temperature and pressure without which it will be necessary to periodically remove the coke from the soaking drum.

Two visbreaking processes are commercially available: the *soaker visbreaker* and the *coil visbreaker*. The difference is that the soaker visbreaker works with a lower temperature with a higher residence time, and the coil visbreaker operates at a higher temperature with a lower residence time. The end result is that the soaker visbreaker has lower energy consumption for the same visbreaking severity.

The *soaker visbreaking process* achieves some conversion within the heater, but the majority of the conversion occurs in a reaction vessel or soaker that holds the two-phase effluent at an elevated temperature for a predetermined length of time. Soaker visbreaking is described as a low temperature, high residence time route. Product quality and yields from the coil and soaker drum design are essentially the same at a specified severity, being independent of visbreaker configuration. By providing the residence time required to achieve the desired reaction, the soaker drum design allows the heater to operate at a lower outlet temperature (thereby saving fuel), but there are disadvantages. The main disadvantage is the decoking operation of the heater and the soaker drum and, although decoking requirements of the soaker drum design are not as frequent as those of the coil-type design, the soaker design requires more equipment for coke removal and handling. The customary practice of removing coke from a drum is to cut it out with high-pressure water thereby producing a significant amount of coke-laden water that needs to be handled, filtered, and then recycled for use again.

The *coil visbreaking process* differs from soaker visbreaking insofar as the coil process achieves conversion by high-temperature cracking within a dedicated soaking coil in the furnace. With conversion primarily achieved as a result of temperature and residence time, coil

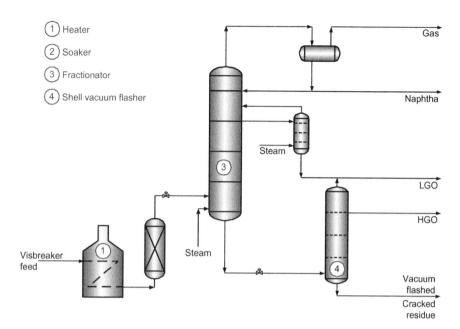

1. Heater
2. Soaker
3. Fractionator
4. Shell vacuum flasher

Gas

Naphtha

Steam

LGO

HGO

Steam

Visbreaker feed

Vacuum flashed

Cracked residue

Figure 2.2 Shell soaker visbreaking technology.

visbreaking is described as a high-temperature, short residence time route. The main advantage of the coil-type design is the two-zone fired heater that provides better control of the material being heated and, with the coil-type design, decoking of the heater tubes is accomplished more easily by the use of steam−air decoking.

In the Shell soaker visbreaking process (Figure 2.2), the feedstock is pumped through preheat exchangers before entering the visbreaker heater, where the residue is heated to the required cracking temperature. In the convection section of the visbreaker heater, superheated steam is generated. Heater effluent is sent to the soaker drum where most of the thermal cracking and viscosity reduction takes place under controlled conditions. The pressure in the soaker drum can be adjusted, which results in a change in residence time and the amount of heavies that reside in the liquid phase thereby providing the possibility to reach optimum selectivity. Soaker drum effluent is flashed and then quenched in the fractionator and the flashed vapors are fractionated into gas, naphtha, gas oil, and visbreaker residue—as anticipated, product yields are dependent on feed type and product specifications. The visbreaker residue is steam-stripped in the bottom of the fractionator and pumped through the cooling circuit to battery limits. Visbreaker gas oil, which is drawn off as a sidestream, is steam-stripped, cooled, and sent to battery limits. Alternately, the gas oil fraction can be included with the visbreaker residue as cutter stock for transportation or catalytic cracker feedstock purposes. The heavy gas oil stream for the visbreaker can be used as feedstock for a thermal distillate cracking unit or for a catalytic cracker for the production of lower boiling distillate products.

Visbreaking has much potential and, in fact, remains an important, relatively inexpensive bottom-of-the-barrel upgrading process in many areas of the world. Most of the existing visbreakers are the soaker type, which utilizes a soaker drum in conjunction with a fired heater to achieve conversion and which reduces the temperature required to achieve conversion while producing a stable residue product, thereby increasing the heater run length and reducing the frequency of unit shut down for heater decoking.

However, a recurring issue with the soaker visbreaker is the need to periodically remove coke from the soaker drum and the inability of the soaker process to easily adjust to changes in feedstock quality because of the need to fine tune two process variables, temperature and residence time. Recent combination of the visbreaking technology and the addition of new coil visbreaker design features have provided the coil process with a competitive advantage over the traditional soaker visbreaker process. Limitations in heater run length are no longer a problem for the coil visbreaker. Advances in visbreaker coil heater design now allow for the isolation of one or more heater passes for decoking, eliminating the need to shut the entire visbreaker down for furnace decoking.

The higher heater outlet temperature specified for a coil visbreaker is now viewed as an important advantage of coil visbreaking. The higher heater outlet temperature is used to recover significantly higher quantities of heavy visbroken gas oil—capability cannot be achieved with a soaker visbreaker without the addition of a vacuum flasher.

2.2.2 Deep Thermal Conversion Process
The *deep thermal conversion* (DTC) process offers a bridge between visbreaking and coking and provides maximum distillate yields by applying DTC to vacuum residua followed by vacuum flashing of the products.

In this process, the heated vacuum residuum is charged to the heater and from there to the soaker where conversion occurs. The products are then led to an atmospheric fractionator to produce gases, naphtha, kerosene, and gas oil. The fractionator residuum is sent to a vacuum flasher that recovers additional gas oil and distillate. The next step for the coke is dependent on its potential use, and it may be isolated as *liquid* coke (pitch, cracked residuum) or solid coke.

2.2.3 ET-II Process

The ET-II process is a thermal cracking process that is designed for feedstocks such as heavy oil and residua (Speight and Ozum, 2002). The distillate (referred to in this process as *cracked oil*) is suitable as a feedstock to hydrocracker and fluid catalytic cracking. The basic technology of the ET-II process is derived from that of the original Eureka process.

In the two-stage process, the feedstock is heated up to 350°C (660°F) by passage through the preheater and fed into the base of the fractionator, where it is mixed with recycle oil and the high-boiling fraction of the cracked oil. The ratio of recycle oil to feedstock is within the range of 0.1−0.3% w/w. The feedstock mixed with recycle oil is then pumped out and fed into the cracking heater, where the temperature is raised to approximately 490−495°C (915−925°F) and the outflow is fed to the stirred-tank reactor where it is subjected to further thermal cracking.

The heat required for the cracking reaction is brought in by the effluent itself from the cracking heater, as well as by the superheated steam, which is heated in the convectional section of the cracking heater and blown into the reactor bottom. The superheated steam reduces the partial pressure of the hydrocarbons in the reactor and accelerates the stripping of volatile components from the cracked heavy feedstock. This heavy feedstocks product is discharged through a transfer pump and transferred to a cooling drum, where the thermal cracking reaction is terminated by quenching with a water spray after which it is sent to the pitch water slurry preparation unit.

2.2.4 Eureka Process

The Eureka process is a thermal cracking process that is designed to produce a cracked oil and aromatic residuum from heavy feedstocks (Aiba et al., 1981).

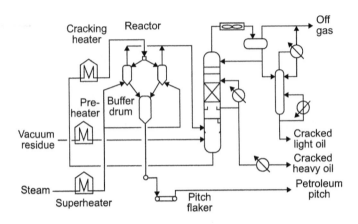

Figure 2.3 The Eureka process.

In this process (Figure 2.3), the heavy feedstock is fed to the preheater and then enters the bottom of the fractionator, where it is mixed with the recycle oil. The feed—recycle oil mixture is then fed to the reactor system, which consists of a pair of reactors operating alternately. In the reactor, a thermal cracking reaction occur in the presence of superheated steam which is injected to strip the cracked products out of the reactor and supply a part of heat required for cracking reaction. At the end of the reaction, the bottom product is quenched. The oil and gas products (and steam) pass from the top of the reactor to the lower section of the fractionator, where a small amount of entrained material is removed by a wash operation. The upper section is an ordinary fractionator, to where the heavier fraction of cracked oil is drawn as a sidestream. The process bottoms (pitch) can be used as boiler fuel, as partial oxidation feedstock for producing hydrogen and carbon monoxide, and as binder pitch for manufacturing metallurgical coke.

The process reactions proceed at lower cracked oil partial pressure by injecting steam into the reactor, and by keeping petroleum pitch in a homogeneous liquid state, unlike with a conventional delayed coker, a higher cracked oil yield can be obtained. After hydrotreating, the cracked oil is used as feedstock for a fluid catalytic cracker or a hydrocracker.

The original Eureka process used two batch reactors, while the later version of the ET-II process and the high conversion soaker cracking (HSC) process both employ continuous reactors.

2.2.5 Fluid Thermal Cracking Process

The fluid thermal cracking (FTC) process is a heavy oil and heavy feedstock upgrading process in which the feedstock is thermally cracked to produce distillate and coke. The coke is gasified to fuel gas (Miyauchi and Ikeda, 1988; Miyauchi et al., 1981, 1987).

The feedstock, mixed with recycle stock from the fractionator, is injected into the cracker and is immediately absorbed into the pores of the particles by capillary force and is subjected to thermal cracking. Hydrogen-containing gas from the fractionator is used for the fluidization in the cracker and excessive coke caused by the metals accumulated on the particle is suppressed under the presence of hydrogen. The particles with deposited coke from the cracker are sent to the gasifier, where the coke is gasified and converted into carbon monoxide (CO), hydrogen (H_2), carbon dioxide (CO_2), and hydrogen sulfide (H_2S) with steam and air. Regenerated hot particles are returned to the cracker.

2.2.6 High Conversion Soaker Cracking Process

The HSC process is a cracking process designed for moderate conversion, higher than visbreaking but lower than coking (Washimi, 1989; Watari et al., 1987). The process features less gas-make and a higher yield of distillate compared to other thermal cracking processes. The process can be used to convert a wide range of feedstocks with high sulfur and metals content, including heavy oils, oil sand bitumen, heavy feedstocks, and visbroken heavy feedstocks. As a note of interest, the HSC process both employs continuous reactors whereas the original Eureka process uses two batch (or semibatch) reactors.

The preheated feedstock enters the bottom of the fractionator, where it is mixed with the recycle oil. The mixture is pumped up to the charge heater and fed to the soaking drum (*ca.* atmospheric pressure, steam injection at the top and bottom), where sufficient residence time is provided to complete the thermal cracking. In the soaking drum, the feedstock and some product flows downward passing through a number of perforated plates while steam with cracked gas and distillate vapors flow through the perforated plates countercurrently.

The volatile products from the soaking drum enter the fractionator where the distillates are fractionated into desired product oil streams, including a heavy gas oil fraction. The cracked gas product is compressed and used as refinery fuel gas after sweetening. The cracked oil

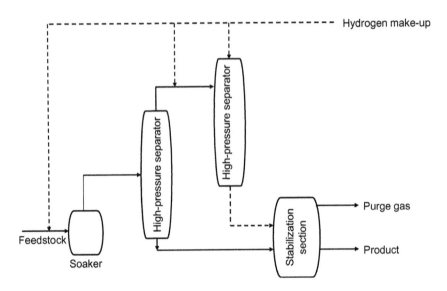

Figure 2.4 The Tervahl process.

product after hydrotreating is used as fluid catalytic cracking or hydro-cracker feedstock. The heavy feedstock is suitable for use as boiler fuel, road asphalt, binder for the coking industry, and as feedstock for partial oxidation.

2.2.7 Tervahl Process
Other variations of visbreaking technology include the Tervahl T and Tervahl H processes (Figure 2.4). The Tervahl T alternative includes only the thermal section to produce a synthetic crude oil with better transportability by having reduced viscosity and greater stability. The Tervahl H alternative adds hydrogen that also increases the extent of the desulfurization and decreases the carbon heavy feedstocks.

In the Tervahl T process (LePage et al., 1987; Peries et al., 1988), the feedstock is heated to the desired temperature using the coil heater and heat recovered in the stabilization section is held for a specified residence time in the soaking drum. The soaking drum effluent is quenched and sent to a conventional stabilizer or fractionator where the products are separated into the desired streams (Table 2.3). The gas produced from the process is used for fuel.

In the related Tervahl H process (a hydrogenation process but cov-ered here for convenient comparison with the Tervahl T process), the

Table 2.3 Feedstock and Product Data for the Tervahl T and Tervahl H Processes

Feedstock: Boscan heavy crude oil		
API	10.5	
Distillate, wt. %		
< 500°C, <930°F	35.5	
Process		
	Tervahl T	Tervahl H
Product		
API	11.7	14.8
Distillate, wt. %		
< 500°C, <930°F	52.5	55.3

feedstock and hydrogen-rich stream are heated using heat recovery techniques and fired heater and held in the soaking drum as in the Tervahl T process. The gas and oil from the soaking drum effluent are mixed with recycle hydrogen and separated in the hot separator where the gas is cooled and passed through a separator and recycled to the heater and soaking drum effluent. The liquids from the hot and cold separator are sent to the stabilizer section where purge gas and synthetic crude are separated. The gas is used as fuel and the synthetic crude can now be transported or stored.

2.3 COKING PROCESSES

Coking processes generally utilize longer reaction times than thermal cracking processes. To accomplish this, drums or chambers (reaction vessels) are employed, but it is necessary to use two or more such vessels so that coke removal can be accomplished in those vessels not onstream without interrupting the semicontinuous nature of the process.

The formation of large quantities of coke is a severe drawback unless the coke can be put to use. Calcined petroleum coke can be used for making anodes for aluminum manufacture and a variety of carbon or graphite products such as brushes for electrical equipment. These applications, however, require a coke that is low in mineral matter and sulfur.

If the feedstock produces a high-sulfur, high-ash, high-vanadium coke, one option for use of the coke is combustion of the coke to produce process steam (and large quantities of sulfur dioxide unless the coke is first gasified or the combustion gases are scrubbed). Another option is stockpiling.

2.3.1 Delayed Coking

Delayed coking is a semicontinuous (semibatch) process in which the heated charge is transferred to large soaking (or coking) drums, which provide the long residence time needed to allow the cracking reactions to proceed to completion (Feintuch and Negin, 1997; McKinney, 1992).

The delayed coking process (Figure 2.5) is widely used for treating heavy feedstocks and uses long reaction times in the liquid phase to convert the residue fraction of the feed to gases, distillates, and coke. The condensation reactions that give rise to the highly aromatic coke product also tend to retain sulfur, nitrogen, and metals, so that the coke is enriched in these elements relative to the feed.

In the process, the feedstock is introduced into the product fractionator where it is heated and lighter fractions are removed as sidestream products. The fractionator bottoms, including a recycle stream of heavy product, are then heated in a furnace whose outlet temperature varies from 480°C to 515°C (895–960°F). The heated feedstock enters one of a pair of coking drums where the cracking reactions continue. The

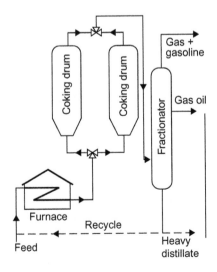

Figure 2.5 A delayed coker.

cracked products leave as an overhead stream and coke deposits on the inner surface of the drum. To give continuous operation, two drums are used; while one is onstream, the other is being cleaned. The temperature in the coke drum ranges from 415°C to 450°C (780–840°F) at pressures from 15 to 90 psi.

Overhead products go to the fractionator where naphtha and heating oil fractions are recovered. The heavy recycle material is combined with preheated fresh feed and returned to the reactor.

A pair of coke drum is used so that while one drum is onstream, the other is being cleaned thereby allowing continuous processing, and the drum operation cycle is typically 48 h. A coke drum is usually onstream for about 24 h before becoming filled with porous coke and the following procedure is used to remove the coke: (i) the coke deposit is cooled with water, (ii) one of the heads of the coking drum is removed to permit the drilling of a hole through the center of the deposit, and (iii) a hydraulic cutting device, which uses multiple high-pressure water jets, is inserted into the hole and the wet coke is removed from the drum. The flexibility of operation inherent in delayed coking permits refiners to process a wide variety of crude oils including those containing heavy, high-sulfur heavy feedstocks (Speight and Ozum, 2002). As expected, yields and product quality vary widely due to the broad range of feedstock types available for delayed coking units, and there is a decrease in overhead yield with asphaltene content of the feedstock (Schabron and Speight, 1997).

The disadvantages of delayed coking are that it is a thermal cracking process, and it is a more expensive process than solvent deasphalting (see Chapter 6), although still less expensive than other conversion processes for heavier crude oil. One common misconception of delayed coking is that the product coke can be a disadvantage. Although coke is a low-valued by-product, compared to transportation fuels, there is a significant worldwide trade and demand even for high-sulfur petroleum coke from delayed cokers as coke is a very economical fuel. However, for coke burning power plants, this has required the installation of flue gas scrubbers and, in some of these plants, a circulating bed of limestone captures the sulfur.

Other similar processes are available and two worthy of mention here are (i) low-pressure coking and (ii) high-temperature coking.

Low-pressure coking is a process designed for a once-through, low-pressure operation. The process is similar to delayed coking except that recycling is not usually practiced and the coke chamber operating conditions are 435°C (815°F), at 25 psi. Excessive coking is inhibited by the addition of water to the feedstock in order to quench and restrict the reactions of the reactive intermediates.

High-temperature coking is a semicontinuous thermal conversion process designed for high-melting asphaltic heavy feedstocks that yields coke and gas oil as the primary products. The coke may be treated to remove sulfur to produce a low-sulfur coke ($\leq 5\%$), even though the feedstock contained as much as 5% w/w sulfur.

Thus, the feedstock is transported to the pitch accumulator, then to the heater (370°C, 700°F, 30 psi), and finally to the coke oven, where temperatures may be as high as 980−1095°C (1800−2000°F). Volatile materials are fractionated, and after the cycle is complete, coke is collected for sulfur removal and quenching before storage.

2.3.2 Fluid Coking

Fluid coking is a continuous process that uses the fluidized solids technique to convert heavy feedstocks, including vacuum heavy feedstocks and cracked heavy feedstocks, to more valuable products (Roundtree, 1997). The process is useful for processing heavy feedstocks. The yield of distillates from coking can be improved by reducing the residence time of the cracked vapors. In order to simplify handling of the coke product, and enhance product yields, fluidized-bed coking or fluid coking was developed in the mid-1950s.

The heavy feedstock is decomposed by being sprayed into a fluidized bed of hot, fine-coke particles, which permits the coking reactions to be conducted at higher temperatures and shorter contact times that can be employed in delayed coking. Moreover, these conditions result in decreased yields of coke; greater quantities of more valuable liquid products are recovered in the fluid coking process.

Fluid coking uses two vessels, a reactor and a burner; coke particles are circulated between these to transfer heat (generated by burning a portion of the coke) to the reactor (Figure 2.6; Blaser, 1992). The reactor holds a bed of fluidized coke particles and steam is introduced at the bottom of the reactor to fluidize the bed. The feed coming from

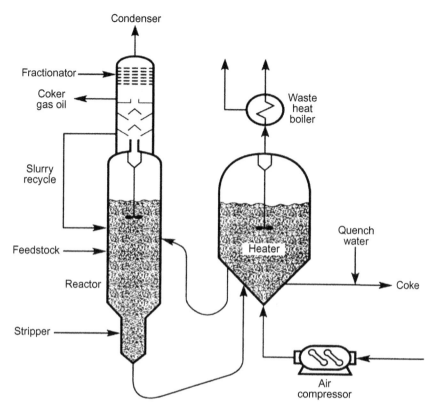

Figure 2.6 A fluid coker.

the bottom of a vacuum tower, for example, 260–370°C (500–700°F), is injected directly into the reactor. The temperature in the coking vessel ranges from 480°C to 565°C (900–1050°F), and the pressure is substantially atmospheric so the incoming feed is partly vaporized and partly deposited on the fluidized coke particles. The material on the particle surface then cracks and vaporizes, leaving a residue that dries to form coke. The vapor products pass through cyclones that remove most of the entrained coke.

The vapor is discharged into the bottom of a scrubber, where the products are cooled to condense a heavy tar containing the remaining coke dust, which is recycled to the coking reactor. The upper part of the scrubber tower is a fractionating zone from which coker gas oil is withdrawn and then fed to a catalytic cracking unit; naphtha and gas are taken overhead to condensers.

In the reactor, the coke particles flow down through the vessel into a stripping zone at the bottom. Steam displaces the product vapors between the particles, and the coke then flows into a riser that leads to the burner. Steam is added to the riser to reduce the solids loading and to induce upward flow. The average bed temperature in the burner is 590–650°C (1095–1200°F), and air is added as needed to maintain the temperature by burning part of the product coke. The pressure in the burner may range from 5 to 25 psi. Flue gases from the burner bed pass through cyclones and discharge to the stack. Hot coke from the bed is returned to the reactor through a second riser assembly.

Coke is one of the products of the process, and it must be withdrawn from the system to keep the solids inventory from increasing. The net coke produced is removed from the burner bed through a quench elutriator drum, where water is added for cooling and cooled coke is withdrawn and sent to storage. During the course of the coking reaction, the particles tend to grow in size. The size of the coke particles remaining in the system is controlled by a grinding system within the reactor.

As expected, the yields of products are determined by the feed properties, the temperature of the fluid bed, and the residence time in the bed. The use of a fluidized bed reduces the residence time of the vapor-phase products in comparison to delayed coking, which in turn reduces cracking reactions. The yield of coke is thereby reduced, and the yield of gas oil and olefins is increased. An increase of 5°C (9°F) in the operating temperature of the fluid-bed reactor typically increases gas yield and naphtha yield by about 1% by weight.

The disadvantage of burning the coke to generate process heat is that sulfur from the coke is liberated as sulfur dioxide. The gas steam from the coke burner also contains carbon monoxide (CO), carbon dioxide (CO_2), and nitrogen (N_2). An alternate approach is to use a coke gasifier to convert the carbonaceous solids to a mixture of carbon monoxide (CO), carbon dioxide (CO_2), and hydrogen (H_2).

Delayed coking and fluid coking are the original processes of choice for conversion of Athabasca bitumen to liquid products. Both processes are termed the *primary conversion* processes for the tar sand plants in Fort McMurray, Canada. The unstable liquid product streams are hydrotreated before recombining to the synthetic crude oil (Speight, 2000, 2007, 2011; Spragins, 1978).

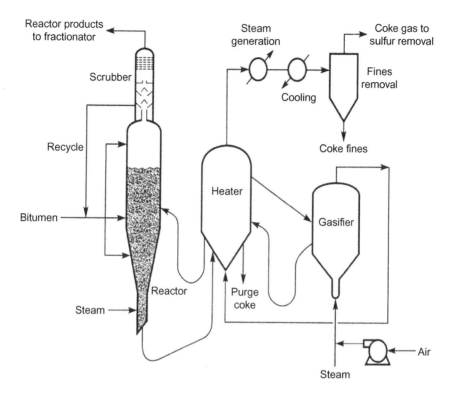

Figure 2.7 Flexicoking process.

2.3.3 Flexicoking

The *flexicoking* process is a direct descendent of fluid coking and uses the same configuration as the fluid coker but includes a gasification section in which excess coke can be gasified to produce refinery fuel gas (Figure 2.7; Roundtree, 1997). The flexicoking process was designed during the late 1960s and 1970s as a means by which excess coke could be reduced in view of the gradual incursion of the heavier feedstocks in refinery operations. Such feedstocks are notorious for producing high yields of coke (>15% by weight) in thermal and catalytic operations.

In this process, excess coke is converted to a low heating value gas in a fluid bed gasifier with steam and air. The air is supplied to the gasifier to maintain temperatures of 830−1000°C (1525−1830°F) but is insufficient to burn all of the coke. Under these reducing conditions, the sulfur in the coke is converted to hydrogen sulfide, which can be scrubbed from the gas prior to combustion. A typical gas product, after removal of hydrogen sulfide, contains carbon monoxide (CO, 18%), carbon dioxide (CO_2, 10%), hydrogen (H_2, 15%), nitrogen (N_2, 51%), water (H_2O, 5%), and methane (CH_4, 1%). The heater is located between the reactor and the gasifier, and it serves to transfer heat between the two vessels.

The heart of the flexicoking process is the fluid coking unit, and the yields of liquid products from flexicoking are the same as from fluid coking. The main difference between fluid coking and flexicoking is the presence of a gasification reactor to assure high conversion of the coke. Units are designed to gasify 60−97% of the coke from the reactor. Even with the gasifier, the product coke will contain more sulfur than the feed, which limits the attractiveness of even the most advanced of coking processes.

2.3.4 Asphalt Coking Technology Process

The asphalt coking technology (ASCOT) process is a heavy feedstocks upgrading process that integrates the delayed coking process and the deep solvent deasphalting process (low-energy deasphalting (LEDA) (Chapter 6); (Bonilla, 1985; Bonilla and Elliott, 1987).

In this process, the heavy feedstock is raised to the desired extraction temperature (50−230°C, 120−445°F, at 300−500 psig, 2060−3430 kPa) and then sent to the extractor where solvent (straight run naphtha, coker naphtha; solvent to oil ratio = 4:1−13:1) flows upward, extracting soluble material from the down-flowing feedstock. The solvent-deasphalted phase leaves the top of the extractor and flows to the solvent recovery system where the solvent is separated from the deasphalted oil and recycled to the extractor.

The deasphalted oil is sent to the delayed coker (heater outlet temperature: 480−510°C, 900−950°F, at 15−35 psig, 105−240 kPa and a recycle ratio of 0.05−0.25 on fresh feedstock) where it is combined with the heavy coker gas oil from the coker fractionator and sent to the heavy coker gas oil stripper where low-boiling hydrocarbons are stripped off and returned to the fractionator. The stripped deasphalted oil/heavy coker gas oil mixture is removed from the bottom of the stripper and is used to provide heat to the naphtha stabilizer−reboiler before being sent to battery limits as a cracking stock. The raffinate phase containing the asphalt and some solvent flowing at a controlled rate from the bottom of the extractor is charged directly to the coking section.

The solvent contained in the asphalt and deasphalted oil is condensed in the fractionator overhead condensers, where it can be recovered and used as lean oil for propane/butane recovery in the absorber, eliminating the need to recirculate lean oil from the naphtha stabilizer. The solvent introduced in the coker heater and coke drums results in a significant reduction in the partial pressure of asphalt feed, compared with a regular delayed coking unit. The low asphalt partial pressure results in low coke and high liquid yields in the coking reaction.

With the ASCOT process, there is a significant reduction in by-product fuel as compared to either solvent deasphalting or delayed coking, and the process can be tailored to process a specific quantity or process to a specific quality of cracking stock.

2.3.5 Cherry-P Process

The Cherry-P process (comprehensive heavy ends reforming refinery process) is a process for the conversion of heavy crude oil or heavy feedstock into distillate and a cracked heavy feedstock (Ueda, 1976, 1978). In this process, the principal aim is to upgrade heavy feedstocks at conditions between those of conventional visbreaking and delayed coking. Although coal is added to the feedstock, it is mainly intended to act as a scavenger to prevent coke buildup on the reactor wall.

In this process, the heavy feedstock is mixed with coal powder in a slurry mixing vessel, heated in the furnace, and fed to the reactor where the feedstock undergoes thermal cracking reactions for several (3−5) hours at a temperature on the order of higher than 400−430°C (750−805°F) and under pressure (140−280 psig).

The residence time is on the order of 1−5 h. No catalyst or hydrogen is added. Gas and distillate from the reactor are sent to a fractionator and the cracked heavy feedstock residue is extracted out of the system after distilling low-boiling fractions by the flash drum and vacuum flasher to adjust its softening point. Distillable product yields of 44% by weight on total feed are reported. Since this yield is obtained when using anthracite, the proportion that is derived from the coal is likely to be very low.

The distillates produced by this process are generally lower in the content of olefin hydrocarbons than with the other thermal cracking process, comparatively easy to desulfurize in hydrotreating units, and compatible with straight run distillates.

2.3.6 Decarbonizing

The thermal *decarbonizing* process is designed to minimize coke and gasoline yields but, at the same time, to produce maximum yields of gas oil. Thermal decarbonizing is essentially the same as the delayed coking process but lower temperatures and pressures are employed. For example, pressures range from 10 to 25 psi, heater outlet temperatures may be 485°C (905°F), and coke drum temperatures may be of the order of 415°C (780°F).

Decarbonizing—as used in this current context—should not be confused with *propane decarbonizing*, which is essentially a nonthermal solvent deasphalting process (see Chapter 6).

2.3.7 Continuous Coking Process

A new coking process that can accept heavy feedstock and continuously discharge vapor and dry petroleum coke particles has been developed (Sullivan, 2011). The process promotes a rapid recovery of volatiles from the residuum enabling recovery of more volatiles. It also causes the carbonization reactions to proceed more rapidly, and it produces uniform composition and uniform size of coke particles that have a low volatiles content.

The new process (Figure 2.8) uses a kneading and mixing action to continuously expose new resid surface to the vapor space and causes a more complete removal of volatiles from the produced petroleum coke. Not only are more valuable volatiles recovered but also they are likely to be richer in middle distillates. As a result of kneading/mixing action by the reactor/devolatilizer, new surfaces of the residuum mass are continuously exposed to the gas phase, enhancing the rapid mass transfer of volatiles into the gas phase. The volatiles are then rapidly cooled to retard degradation. With the rapid reduction of volatiles content in the resid mass, the carbonization reaction rates are accelerated, enabling continuous and rapid production of solid petroleum coke particles. The short contact time of the volatiles with the hot residuum minimizes thermal degradation of volatiles.

Concurrently with the carbonization reactions and the formation of coke, some cracking of side chains off the larger molecules likely

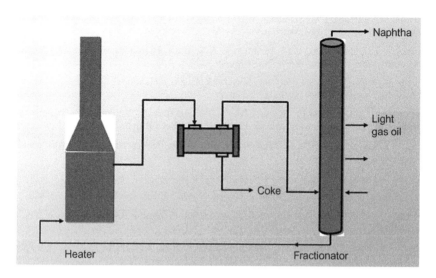

Figure 2.8 Continuous coking process.

occurs. These smaller, low-boiling molecules produced from cracking reactions join the population of the indigenous volatiles. Some volatiles may be generated even after the solid coke is formed. In the delayed coking process, many of these late-forming volatiles remain trapped in the coke. The process promotes the release of these late-forming cracked volatiles, allowing them to escape into the gas phase by breaking the solid coke into small particles.

In addition to the recovery of additional and more valuable volatiles, there are other benefits of the new process compared to delayed coking. The consumption of utilities is less because no steam or water is required. Since there is no quenching, energy from the hot coke is recovered. The process is continuous so is never opened to the atmosphere. There is no cutting procedure as in the delayed coking process where high-pressure water is used to cut the coke out of the drums. Therefore, no volatiles and no coke particles are released into the atmosphere.

REFERENCES

Aiba, T., Kaji, H., Suzuki, T., Wakamatsu, T., 1981. Chem. Eng. Prog. February, 37.

Blaser, D.E., 1992. In: McKetta, J.J. (Ed.), Petroleum Processing Handbook. Marcel Dekker, New York, NY, p. 255.

Bonilla, J.A., 1985. Energy Prog. 5 (4), 239–244 (December).

Bonilla, J.A., Elliott, J.D., 1987. Asphalt Coking Method. United States Patent 4,686,027. August 11.

Dickenson, R.L., Biasca, F.E., Schulman, B.L., Johnson, H.E., 1997. Refiner options for converting and utilizing heavy fuel oil. Hydrocarbon Process. 76 (2), 57−62.

Dominici, V.E., Sieli, G.M., 1997. In: Meyers, R.A. (Ed.), Handbook of Petroleum Refining Processes. McGraw-Hill, New York, NY (Chapter 12.3).

Feintuch, H.M., Negin, K.M., 1997. In: Meyers, R.A. (Ed.), Handbook of Petroleum Refining Processes. McGraw-Hill, New York, NY (Chapter 12.2).

Gary, J.H., Handwerk, G.E., Kaiser, M.J., 2007. Petroleum Refining: Technology and Economics, fifth ed. CRC Press, Taylor & Francis Group, Boca Raton, FL.

Hsu, C.S., Robinson, P.R. (Eds.), 2006. Practical Advances in Petroleum Processing, vols. 1 and 2. Springer Science, New York, NY.

Hulet, C., Briens, C., Berruti, F., Chan, E.W., 2005. Int. J. Reactor Eng. 3, R1.

LePage, J.F., Morel, F., Trassard, A.M., Bousquet, J., 1987. Prepr. Div. Fuel Chem. 32, 470.

McKinney, J.D., 1992. In: McKetta, J.J. (Ed.), Petroleum Processing Handbook. Marcel Dekker, New York, NY, p. 245.

Miyauchi, T., Ikeda, Y., Kikuchi, T. 1988. Process for thermal cracking of heavy oil. US Patent 4,772,378. September 20.

Miyauchi, T., Furusaki, S., Morooka, Y., 1981. Advances in Chemical Engineering. Academic Press, New York, NY (Chapter 11).

Miyauchi, T., Tsutsui, T., Nozaki, Y., 1987. A new fluid thermal cracking process for upgrading resid. Paper 65B. Proceedings of the Spring National Meeting. American Institute of Chemical Engineers, March 29, Houston, TX.

Peries, J.P., Quignard, A., Farjon, C., Laborde, M., 1988. Thermal and catalytic ASVAHL processes under hydrogen pressure for converting heavy crudes and conventional residues. Rev. Inst. Fr. du Pétrole 43 (6), 847−853.

Roundtree, E.M., 1997. In: Meyers, R.A. (Ed.), Handbook of Petroleum Refining Processes. McGraw-Hill, New York, NY (Chapter 12.1).

Schabron, J.F., Speight, J.G., 1997. Rev. Inst. Fr. du Pétrole 52 (1), 73−85.

Speight, J.G., 1990. Fuel Science and Technology Handbook. Marvel Dekker Inc., New York, Chapters 12−16.

Speight, J.G., 2000. The Desulfurization of Heavy Oils and Residua, second ed. Marcel Dekker Inc., New York, NY.

Speight, J.G., 2007. The Chemistry and Technology of Petroleum, fourth ed. CRC Press, Taylor & Francis Group, Boca Raton, FL.

Speight, J.G., 2011. The Refinery of the Future. Gulf Professional Publishing, Elsevier, Oxford, UK.

Speight, J.G., Ozum, B., 2002. Petroleum Refining Processes. Marcel Dekker, New York, NY.

Spragins, F.K., 1978. In: Chilingarian, G.V., Yen, T.F. (Eds.), Bitumens, Asphalts, and Tar Sands. Elsevier, Amsterdam, the Netherlands, p. 92.

Sullivan, D.W., 2011. New continuous coking process. Proceedings of the 14th Topical Symposium on Refinery Processing. AIChE Spring Meeting and Global Congress on Process Safety, March 13−17, Chicago, IL.

Ueda, H., 1978. J. Fuel Soc. Jpn. 57, 963.

Ueda, K., 1976. J. Jpn. Pet. Inst. 19 (5), 417.

Washimi, K., 1989. Hydrocarbon Process. 68 (9), 69.

Watari, R., Shoji, Y., Ishikawa, T., Hirotani, H., Takeuchi, T., 1987. Annual Meeting. National Petroleum Refiners Association, San Antonio, TX. Paper AM-87-43.

CHAPTER 3

Catalytic Cracking

3.1 INTRODUCTION

Catalytic cracking has a number of advantages over thermal cracking (Table 3.1; Avidan and Krambeck, 1990; Gary et al., 2007; Hsu and Robinson, 2006; Speight, 2007, 2011; Speight and Ozum, 2002). Typically, the feedstocks for catalytic cracking can be any one (or blends) of the following: (i) straight-run gas oil, (ii) vacuum gas oil, and (iii) heavy feedstocks such as heavy oil, atmospheric residuum, vacuum residuum, extra-heavy oil, and tar sand bitumen.

However, if blends of the above feedstocks are employed, compatibility of the constituents of the blends must be assured under read to conditions or excessive coke will be laid down on to the catalyst (Speight, 2007, 2011). In addition, there are several pretreatment options for the feedstocks for catalytic cracking units and these are: (i) deasphalting to prevent excessive coking on catalyst surfaces, (ii) demetallization, that is, removal of nickel, vanadium, and iron to prevent catalyst deactivation, (iii) use of a short residence time as a means of preparing the feedstock, and (iv) hydrotreating or mild hydrocracking to prevent excessive coking in the fluid catalytic cracking (FCC) unit (Bartholic, 1981a,b; Speight, 2000, 2004; Speight and Ozum, 2002).

Hydrotreating the fluid catalytic cracker feed improves naphtha yield (Table 3.2) and quality and reduces the sulfur oxide (SO$_x$) emissions from the catalytic cracker unit, but it is typically a high-pressure process and, furthermore, manipulation of feedstock sulfur alone may not be sufficient to meet future gasoline performance standards. Refineries wishing to process heavier crude oil may only have the option to desulfurize the resulting higher sulfur naphtha. Hydrodesulfurization (HDS) of catalytic cracker naphtha is a low-pressure process. Obviously, the selection of an optimum hydrotreating process option for reducing sulfur in catalytic cracker naphtha is determined by economic factors specific to a refinery and to the feedstock. Hydrotreating catalytic cracker feedstock can be very profitable for a

Table 3.1 Comparison of Thermal Cracking and Catalytic Cracking

Thermal cracking	Catalytic cracking
No catalyst	Uses a catalyst
Higher temperature	Lower temperature
Higher pressure	Lower pressure
Free radical reaction mechanisms	More flexible in terms of product slate
Moderate thermal efficiency	Ionic reaction mechanisms
No regeneration of catalyst needed	High thermal efficiency
Moderate yields of gasoline and other distillates	Good integration of cracking and regeneration
Gas yields feedstock dependent	High yields of gasoline and other distillates
Low-to-moderate product selectivity	Low gas yields
Alkanes produced but feedstock-dependent yields	High product selectivity
Low octane number gasoline	Low n-alkane yields
Some chain branching in alkanes	High octane number
Low-to-moderate yield of C_4 olefins	Chain branching and high yield of C_4 olefins
Low-to-moderate yield of aromatics	High yields of aromatics

Table 3.2 Feedstock and Product Data for the Fluid Catalytic Process with and Without Feedstock Hydrotreating

Feedstock ($>370°C$, $>700°F$)	No Pretreatment	With Hydrotreatment
API	15.1	20.1
Sulfur, wt%	3.3	0.5
Nitrogen, wt%	0.2	0.1
Carbon residue, wt%	8.9	4.9
Nickel + vanadium, ppm	51.0	7.0
Products		
Naphtha (C5—221°C, C5—430°F), vol.%	50.6	58.0
Light cycle oil (221−360°C, 430−680°F), vol.%	21.4	18.2
Residuum (>360°C, >680°F) wt%	9.7	7.2
Coke, wt%	10.3	7.0

refiner despite the large capital investment involved. By taking advantage of feedstock hydrotreating, margins can be optimized by considering the feedstock hydrotreater unit, the fluid catalytic cracker, and any postcatalytic cracker hydrotreaters as one integrated upgrading step.

Catalytic cracking in the usual commercial process involves contacting the feedstock with a catalyst under suitable conditions of temperature, pressure, and residence time. By this means, a substantial part (>50%) of the feedstock is converted into gasoline and lower boiling products, usually in a single pass. However, during the cracking reaction, carbonaceous material is deposited on the catalyst, which markedly reduces its activity, and removal of the deposit is very necessary. The carbonaceous deposit arises from the thermal decomposition of high-molecular-weight polar species (Speight, 2007) in the feedstock. Removal of the deposit from the catalyst is usually accomplished by burning in the presence of air until catalyst activity is reestablished.

3.2 PROCESS TYPES

Catalytic cracking is another innovation that truly belongs to the twentieth century. It is the modern method for converting high-boiling petroleum fractions, such as gas oil, into gasoline and other low-boiling fractions. The several processes currently employed in catalytic cracking differ mainly in the method of catalyst handling, although there is an overlap with regard to catalyst type and the nature of the products. The catalyst, which may be an activated natural or synthetic material, is employed in bead, pellet, or microspherical form and can be used as a *fixed-bed*, *moving-bed*, or *fluid-bed* configuration.

The *fixed-bed process* was the first to be used commercially and uses a static bed of catalyst in several reactors, which allows a continuous flow of feedstock to be maintained. Thus, the cycle of operations consists of (i) flow of feedstock through the catalyst bed, (ii) discontinuance of feedstock flow and removal of coke from the catalyst by burning, and (iii) insertion of the reactor onstream.

The *moving-bed process* uses a reaction vessel in which cracking takes place and a kiln in which the spent catalyst is regenerated; catalyst movement between the vessels is provided by various means.

The *fluid-bed process* differs from the fixed-bed and moving-bed processes insofar as the powdered catalyst is circulated essentially as a fluid with the feedstock. The several FCC processes in use differ primarily in mechanical design (Hsu and Robinson, 2006; Speight and Ozum, 2002). Side-by-side reactor-regenerator configuration or the reactor either above or below the regenerator is the main mechanical variation. From a flow standpoint, all FCC processes contact the feedstock and any recycle streams with the finely divided catalyst in the reactor.

Feedstocks may range from naphtha to atmospheric residuum (*reduced crude*). Feed preparation (to remove *metallic constituents* and *high-molecular-weight nonvolatile materials*) is usually carried out via any one of the following means: coking, propane deasphalting, furfural extraction, vacuum distillation, viscosity breaking, thermal cracking, and HDS (Speight, 2000).

The major process variables are temperature, pressure, catalyst–feedstock ratio (ratio of the weight of catalyst entering the reactor per hour to the weight of feedstock charged per hour), and space velocity (weight or volume of the feedstock charged per hour per weight or volume of catalyst in the reaction zone). Wide flexibility in product distribution and quality is possible through control of these variables along with the extent of internal recycling is necessary. Increased conversion can be obtained by applying higher temperature or higher pressure. Alternatively, lower space velocity and higher catalyst–feedstock ratio will also contribute to an increased conversion.

When cracking is conducted in a single stage, the more reactive hydrocarbons may be cracked, with a high conversion to gas and coke, in the reaction time necessary for reasonable conversion of the more refractory hydrocarbons. However, in a two-stage process, gas and gasoline from a short reaction time, high-temperature cracking operation are separated before the main cracking reactions take place in a second-stage reactor.

3.2.1 Fixed-Bed Processes

Historically, the Houdry fixed-bed process, which went onstream in June 1936, was the first of the modern catalytic cracking processes. It was preceded only by the McAfee batch process, which employed a metal halide (aluminum chloride) catalyst but has long since lost any commercial significance.

In the fixed-bed process, the catalyst in the form of small lumps or pellets was made up in layers or beds in several (four or more) catalyst-containing drums called converters. Feedstock vaporized at about 450°C (840°F) and less than 7–15 psi pressure passed through one of the converters where the cracking reactions took place. After a short time, deposition of coke on the catalyst rendered it ineffective, and using a synchronized valve system, the feed stream was turned into a neighboring converter while the catalyst in the first converter was regenerated by carefully burning the coke deposits with air. After about 10 min, the catalyst was ready to go onstream again.

Fixed-bed processes have now generally been replaced by moving-bed or fluid-bed processes.

3.2.2 Fluid-Bed Processes

The FCC process (Figure 3.1) is the most widely used process and is characterized by the use of a finely powdered catalyst that is moved through the reactor (Figure 3.2) and flow patterns may vary depending upon the precise configuration of the reactor. The catalyst particles are of such a size that when *aerated* with air or hydrocarbon vapor, the catalyst behaves like a liquid and can be moved through pipes.

Since the catalyst in the reactor becomes contaminated with coke, the catalyst is continuously withdrawn from the bottom of the reactor and lifted by means of a stream of air into a regenerator where the coke is removed by controlled burning. The regenerated catalyst then flows to the fresh feed line, where the heat in the catalyst is sufficient to vaporize the fresh feed before it reaches the reactor, where the temperature is about 510°C (950°F).

Two-stage FCC was devised to permit greater flexibility in shifting product distribution when dictated by demand. Thus, feedstock is first

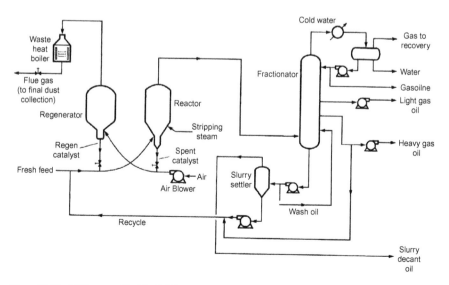

Figure 3.1 The FCC process (Speight, 2007).

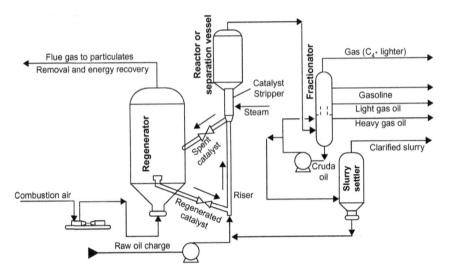

Figure 3.2 Detailed schematic of an FCC reactor.

contacted with cracking catalyst in a riser reactor, that is, a pipe in which fluidized catalyst and vaporized feedstock flow concurrently upward, and the total contact time in this first stage is on the order of seconds. High temperatures, 470−565°C (875−1050°F), are employed to reduce undesirable coke deposits on catalyst without destruction of gasoline by secondary cracking. Other operating conditions in the first stage are a pressure of 16 psi and a catalyst−feedstock ratio of 3:1−50:1, and volume conversion ranges between 20% and 70% have been recorded.

All or part of the unconverted or partially converted gas oil product from the first stage is then cracked further in the second-stage fluid-bed reactor. Operating conditions are 480−540°C (900−1000°F) and 16 psi with a catalyst−oil ratio of 2:1−12:1. Conversion in the second stage varies between 15% and 70%, with an overall conversion range of 50−80%.

3.2.3 Moving-Bed Processes
Residua, ranging from atmospheric residue to vacuum residua including residua high in sulfur or nitrogen, can be used as the feedstock and the catalyst is synthetic or natural. Although the equipment employed is similar in many respects to that used in Houdriflow units, novel process features modify or eliminate the adverse effects on catalyst and

product selectivity usually resulting when metals (such as nickel, vanadium, copper, and iron) are present in the fuel. The Houdresid catalytic reactor and catalyst-regenerating kiln are contained in a single vessel. Fresh feed plus recycled gas oil are charged to the top of the unit in a partially vaporized state and mixed with steam.

The reactor feed and catalyst pass concurrently through the reactor zone to a disengager section, in which vapors are separated and directed to a conventional fractionation system. The spent catalyst, which has been steam purged of residual oil, flows to the kiln for regeneration, after which steam and flue gas are used to transport the catalyst to the reactor.

3.3 PROCESS PARAMETERS

Catalytic cracking is *endothermic* and, that being the case, heat is absorbed by the reactions and the temperature of reaction mixture declines as the reactions proceed and a source of heat for the process is required. This heat comes from combustion of coke formed in the process. Coke is one of the important, though undesirable, products of cracking since it forms on the surface and in the pores of the catalyst during the cracking process, covering active sites and deactivating the catalysts. During regeneration, this coke is burned off the catalyst to restore catalytic activity and, like all combustion processes, the process is *exothermic*, liberating heat.

Most FCC units are operated to maximize conversion to gasoline and liquefied petroleum gas (LPG). This is particularly true when building gasoline inventory for peak season demand or reducing clarified oil yield due to low market demand. Maximum conversion of a specific feedstock is usually limited by both the FCC unit design constraints (i.e., regenerator temperature and wet gas capacity) and the processing objectives. However, within these limitations, the FCC unit operator has many operating and catalyst property variables to select from to achieve maximum conversion.

Each FCC unit that is operated for maximum conversion at constant fresh feed quality has an optimum conversion point beyond which a further increase in conversion reduces the yield of naphtha (gasoline constituents) and increases the yield of LPG and the *optimum conversion point* is referred to as the *overcracking point*.

3.3.1 Process Variables

The primary variables available to the operation of catalytic cracking units for maximum unit conversion for a given feedstock quality can be divided into two groups: (i) process variables and (ii) catalyst variables. In addition to the catalyst variables (*q.v.*), there are also process variables that include (i) pressure, (ii) reaction time, and (iii) reactor temperature. Higher conversion and coke yield are thermodynamically favored by higher *pressure*. However, pressure is usually varied over a very narrow range due to limited air blower horsepower. Conversion is not significantly affected by unit pressure since a substantial increase in pressure is required to significantly increase conversion.

An increase in *reaction time* available for cracking also increases conversion. Fresh feed rate, riser steam rate, recycle rate, and pressure are the primary operating variables which affect reaction time for a given unit configuration. Conversion varies inversely with these stream rates due to limited reactor size available for cracking. Conversion has been increased by a decrease in rate of injection of fresh feedstock. Under these circumstances, overcracking of gasoline to LPG and to dry gas may occur due to the increase in reactor residence time. One approach to offset any potential gasoline overcracking is to add additional riser steam to lower hydrocarbon partial pressure for more selective cracking. Alternatively, an operator may choose to lower reactor pressure or increase the recycle rate to decrease residence time. Gasoline overcracking may be controlled by reducing the availability of catalytic cracking sites by lowering the catalyst—oil ratio.

Increased *reactor temperature* increases feedstock conversion, primarily through a higher rate of reaction for the endothermic cracking reaction and also through increased catalyst—oil ratio. A 10°F increase in reactor temperature can increase conversion by 1—2% absolute but, again, this is feedstock dependent. Higher reactor temperature also increases the amount of olefins in gasoline and in the gases. This is due to the higher rate of primary cracking reactions relative to secondary hydrogen transfer reactions which saturate olefins in the gasoline boiling range and lowers gasoline octane.

However, these variables are not always available for maximizing conversion since most FCC units operate at an optimum conversion level corresponding to a given feed rate, feed quality, set of processing objectives, and catalyst at one or more unit constraints (e.g., wet gas compressor capacity, fractionation capacity, air blower capacity, reactor temperature, regenerator temperature, and catalyst circulation). Once the optimum conversion level is found, there are very few additional degrees of freedom for changing the operating variables.

3.3.2 The Reactor
The three main components of an FCC unit are (i) the reactor, (ii) the stripper, and (iii) the regenerator.

In the unit, the catalyst and the feed and product hydrocarbons are lifted up the riser pipe to the *reactor* where the predominately endothermic cracking processes take place. Since the reactions are endothermic, reaction temperature declines from bottom to top. At the top, the mixture enters a solid–gas separator, and the product vapors are led away. Cracked gases are separated and fractionated; the catalyst and residue, together with recycle oil from a second-stage fractionator, pass to the main reactor for further cracking. The products of this second-stage reaction are gas, gasoline and gas oil streams, and recycle oil.

The coked catalyst enters the *stripper* where steam is added and unreacted–reacted hydrocarbons adsorbed on the catalyst are released. The stripped catalyst is then directed into the *regenerator* where air is added and the combustion of coke on the catalyst (and any hydrocarbons still adsorbed which were not stripped) occurs with the liberation of heat. Regenerator temperatures are typically 705–760°C (1300–1400°F). Heat exchangers and the circulating catalyst capture the heat evolved during regeneration to be used in preheating the reactor feed to appropriate cracking temperatures usually in the range 495–550°C (925–1020°F).

During operations, the entire catalyst inventory is *continually* circulated through the unit. Catalyst residence time in the riser reactor section is typically 1–3 s (with current trends to even shorter residence times), and the entire reactor–stripper–regenerator cycle is less than 10 min. To achieve cycle times of this order, catalyst circulation rates as high as 1 ton/s in large units are required. To withstand such movement, the catalyst must be sufficiently *robust* to withstand the operational stress.

Process temperatures are high, coke is repeatedly deposited and burned off, and the catalyst particles are moving at high speed through steel reactors and pipes. Contact between the catalyst particles and the metal walls and interparticle contact are unavoidable. Thus, catalyst loss from the unit caused by poor attrition resistance can be a serious problem, since the quantities lost must be replaced by fresh catalyst additions to maintain constant unit performance. Catalyst manufacturers work hard to prevent inordinate losses due to attrition and refineries keep a close watch on catalyst quality to be sure the produce conforms to their specifications. Therefore, the robustness of the catalyst is carefully monitored and controlled to a high attrition resistance that is determined by rigorous test methods that place a semiquantitative evaluation on attrition resistance, which is generally related to breakdown with time in commercial units.

As discussed earlier, in some units cracking does not always take place in the reactor and reaction often occurs in the vertical or upward sloped pipe called the *riser* (giving credence to the name *riser reactor* and *riser pipe cracking*) forming products, including coke (Bartholic, 1989). Preheated feedstock is sprayed into the base of the riser via feed nozzles where it contacts extremely hot fluidized catalyst at 1230–1400°F (665–760°C). The hot catalyst vaporizes the feed and catalyzes the cracking reactions that break down the high-molecular-weight oil into lighter components including LPG constituents, gasoline, and diesel. The catalyst–hydrocarbon mixture flows upward through the riser for just a few seconds and then the mixture is separated via cyclones. The catalyst-free hydrocarbons are routed to a main fractionator for separation into fuel gas, propane and butanes, gasoline, light cycle oils used in diesel and jet fuel, and heavy fuel oil.

3.3.3 Coke Formation

The formation of coke deposits has been observed in virtually every unit in the operation, and the deposits can be very thick with thickness up to 4 ft having been reported (McPherson, 1984). Coke has been observed to form where condensation of hydrocarbon vapors occurs. The reactor walls and plenum offer a colder surface where hydrocarbons can condense. Higher boiling constituents in the feedstock may be very close to their dew point, and they will readily condense and form coke nucleation sites on even slightly cooler surfaces.

Unvaporized feed droplets readily collect to form coke precursors on any available surface since the high-boiling feedstock constituents do not vaporize at the mixing zone of the riser. Thus, it is not surprising that residuum processing makes this problem even worse. Low residence time cracking also contributes to coke deposits since there is less time for heat to transfer to feed droplets and vaporize them. This is an observation in line with the increase in coking when short contact time *riser crackers* (*q.v.*) were replacing the longer residence time fluid-bed reactors.

Higher boiling feedstocks that have high aromaticity result in higher yields of coke. Furthermore, polynuclear aromatics and aromatics containing heteroatoms (i.e., nitrogen, oxygen, and sulfur) are more facile coke makers than simpler aromatics (Hsu and Robinson, 2006; Speight, 2007). However, feed quality alone is not a foolproof method of predicting where coking will occur. However, it is known that feedstock hydrotreaters rarely have coking problems. The hydrotreating step mitigates the effect of the coke formers and coke formation is diminished.

It is recognized that significant post-riser cracking occurs in commercial catalytic cracking units resulting in substantial production of dry gas and other low-valued products (Avidan and Krambeck, 1990). There are two mechanisms by which this post-riser cracking occurs, thermal and dilute phase catalytic cracking.

Thermal cracking results from extended residence times of hydrocarbon vapors in the reactor disengaging area and leads to high dry gas yields via nonselective free radical cracking mechanisms. On the other hand, dilute phase catalytic cracking results from extended contact between catalyst and hydrocarbon vapors downstream of the riser. While much of this undesirable cracking was eliminated in the transition from bed to riser cracking, there is still a substantial amount of nonselective cracking occurring in the dilute phase due to the significant catalyst holdup.

Many catalytic cracking units are equipped with advanced riser termination systems to minimize post-riser cracking (Long et al., 1993). However, due to the complexity and diversity of catalytic cracking units, there are many variations of these systems such as closed cyclones and many designs are specific to the unit configuration but all serve the same fundamental purpose of reducing the undesirable post-riser reactions. Furthermore, there are many options for taking advantage of reduced post-riser cracking to improve yields. A combination of higher reactor temperature, higher catalyst−oil ratio, higher feed rate, and/or poorer quality feed is typically employed. Catalyst modification is also appropriate and typical catalyst objectives such as low coke and dry gas selectivity are reduced in importance due to the process changes, while other features such as activity stability and bottoms cracking selectivity become more important for the new unit constraints.

Certain catalyst types seem to increase coke deposit formation. For example, these catalysts (some rare earth zeolites) tend to form aromatics from naphthenes as a result of secondary hydrogen transfer reactions, and the catalysts contribute to coke formation indirectly because the products that they produce have a greater tendency to be coke precursors. In addition, high zeolite content, low surface area cracking catalysts are less efficient at heavy oil cracking than many amorphous catalysts because the nonzeolite catalysts contained a matrix which was better able to crack heavy oils and convert the coke precursors. The active matrix of some modern catalysts serves the same function.

Once coke is formed, it is matter of where it will appear. Coke deposits are most often found in the reactor (or disengager), transfer line, and slurry circuit and cause major problems in some units such as increased pressure drops, when a layer of coke reduces the flow through a pipe, or plugging, when chunks of coke spall off and block the flow completely. Deposited coke is commonly observed in the reactor as a black deposit on the surface of the cyclone barrels, reactor dome, and walls. Coke is also often deposited on the cyclone barrels 180° away from the inlet. Coking within the cyclones can be potentially very troublesome since any coke spalls going down into the dipleg could restrict catalyst flow or jam the flapper valve. Either situation reduces cyclone efficiency and can increase catalyst losses from the reactor. Coke formation also occurs at nozzles which can increase the nozzle pressure drop. It is possible for steam or instrument nozzles to be plugged completely, a serious problem in the case of unit instrumentation.

Coking in the transfer line between the reactor and the main fractionator is also common, especially at the elbow where the line enters the fractionator. Transfer line coking causes pressure drop and spalling and can lead to reduced throughput. Furthermore, any coke in the transfer line which spalls off can pass through the fractionator into the circulating slurry system where it is likely to plug up exchangers, resulting in lower slurry circulation rates and reduced heat removal. Pressure balance is obviously affected if the reactor has to be run at higher pressures to compensate for transfer line coking. On units where circulation is limited by low slide valve differentials, coke laydown may then indirectly reduce catalyst circulation. The risk of a flow reversal is also increased. In units with reactor grids, coking increases grid pressure drop, which can directly affect the catalyst circulation rate.

Shutdowns and start-ups can aggravate problems due to coking. The thermal cycling leads to differential expansion and contraction between the coke and the metal wall that will often cause the coke to spall in large pieces. Another hazard during shutdowns is the possibility of an internal fire when the unit is opened up to the atmosphere. Proper shutdown procedures which ensure that the internals have sufficiently cooled before air enters the reactor will eliminate this problem. In fact, the only defense against having coke plugging problems during start-up is to thoroughly clean the unit during the turnaround and remove all the coke. If strainers are on the line(s), they will have to be cleaned frequently.

The two basic principles to minimize coking are to avoid dead spots and prevent heat losses. An example of minimizing *dead spots* is using purge steam to sweep out stagnant areas in the disengager system. The steam prevents collection of high-boiling condensable products in the cooler regions. Steam also provides a reduced partial pressure or steam distillation effect on the high-boiling constituents and causes enhanced vaporization at lower temperatures. Steam for purging should preferably be superheated since medium-pressure low-velocity steam in small pipes with high heat losses is likely to be very wet at the point of injection and will cause more problems. *Cold spots* are often caused by heat loss through the walls in which case increased thermal resistance might help reduce coking. The transfer line, being a common source of coke deposits, should be as heavily insulated as possible, provided that stress-related problems have been taken into consideration.

In some cases, changing catalyst type or the use of an additive (*q.v.*) can alleviate coking problems. The catalyst types which appear to result in the least coke formation (not delta coke or catalytic coke) contain low or zero earth zeolites with moderate matrix activities. Eliminating heavy recycle streams can lead to reduced coke formation. Since clarified oil is a desirable feedstock to make needle coke in a coker, then it must also be a potential coke maker in the disengager.

One of the trends in recent years has been to improve product yields by means of better feed atomization. The ultimate objective is to produce an oil droplet small enough so that a single particle of catalyst will have sufficient energy to vaporize it. This has the double benefit of improving cracking selectivity and reducing the number of liquid droplets which can collect to form coke nucleation sites.

3.3.4 Additives

In addition to the cracking catalyst described earlier, a series of *additives* have been developed that catalyze or otherwise alter the primary catalyst's activity/selectivity or act as pollution control agents. Additives are most often prepared in microspherical form to be compatible with the primary catalysts and are available separately in compositions that (i) enhance gasoline octane and light olefin formation, (ii) selectively crack heavy cycle oil, (iii) passivate vanadium and nickel present in many heavy feedstocks, (iv) oxidize coke to carbon dioxide, and (v) reduce sulfur dioxide emissions.

Both vanadium and nickel deposit on the cracking catalyst and are extremely deleterious when present in excess of 3000 ppm on the catalyst. Formulation changes to the catalyst can improve tolerance to vanadium and nickel, but the use of additives that specifically passivate either metal is often preferred.

3.4 CATALYSTS AND CATALYST TREATMENT

Cracking heavy oil feedstocks occurs over many types of catalytic materials but high yields of deposited coke are also an undesirable product. To counteract this phenomenon, catalyst activity for the feedstock constituents may be enhanced to some extent by the incorporation of small amounts of other materials, such as the oxides of zirconium (zirconia, ZrO_2), boron (boria, B_2O_3, which has a tendency to volatilize away on use), and thorium (thoria, ThO_2). However, the catalysts must be stable to physical impact loading and thermal shocks and must withstand the action of carbon dioxide, air, nitrogen compounds, and steam. They should also be resistant to sulfur compounds; the synthetic catalysts and certain selected clays appear to be better in this regard than average untreated natural catalysts.

Neither silica (SiO_2) nor alumina (Al_2O_3) alone is effective in promoting catalytic cracking reactions. In fact, they (and also activated carbon) promote decomposition of hydrocarbon constituents that match the thermal decomposition patterns. Mixtures of anhydrous silica and alumina ($SiO_2 \cdot Al_2O_3$) or anhydrous silica with hydrated alumina ($2SiO_2 \cdot 2Al_2O_3 \cdot 6H_2O$) are also essentially not effective. A catalyst having appreciable cracking activity is obtained only when prepared from hydrous oxides followed by partial dehydration (*calcining*). The small amount of water remaining is necessary for proper functioning.

The catalysts are porous and highly adsorptive, and their performance is affected markedly by the method of preparation. Two catalysts that are chemically identical but have pores of different size and distribution may have different activity, selectivity, temperature coefficient of reaction rate, and response to poisons. The intrinsic chemistry and catalytic action of a surface may be independent of pore size, but small pores appear to produce different effects because of the manner and time in which hydrocarbon vapors are transported into and out of the interstices.

The catalyst—oil volume ratios range from 5:1 to 30:1 for the different processes, although most processes are operated to 10:1. However, for moving-bed processes, the catalyst—oil volume ratios may be substantially lower than 10:1.

3.4.1 Catalyst Variables

The primary variables available to the operation of FCC units for maximum unit conversion for a given feedstock quality include catalytic variables such as: (i) catalyst activity and (ii) catalyst design, which includes availability of cracking sites and the presence of carbon on the regenerated catalyst.

The equilibrium *catalyst activity*, as measured by a microactivity test (MAT), is a measure of the availability of zeolite and active matrix cracking sites for conversion. Therefore, an increase in the unit activity can affect an increase in conversion and activity is increased by one, or a combination of: (i) increased fresh catalyst addition rate, (ii) increased fresh catalyst zeolite activity, (iii) increased fresh catalyst matrix activity, (iv) addition of catalyst additives to trap or passivate the deleterious effects of feed nitrogen, alkalis (i.e., calcium and sodium), vanadium, and other feed metal contaminants, and (v) increased fresh catalyst matrix surface area to trap or remove feedstock contaminants.

In general, a two-digit increase in the activity as determined by the MAT appears to coincide with a 1% absolute increase in conversion. The increased matrix surface area improves conversion by providing more amorphous sites for cracking high-boiling range compounds in the feedstock which cannot be cracked by the zeolite. Increased zeolite, on the other hand, provides the necessary acid cracking sites for selectively cracking the amorphous cracked high-boiling compounds and lighter boiling compounds.

In addition to zeolite and matrix activity, many of the catalyst's physical and chemical properties (*catalyst design*) contribute to increased conversion through selectivity differences. These include zeolite type, pore size distribution, relative matrix to total surface area, and chemical composition.

Increasing the concentration of catalyst in the reactor, often referred to as catalyst–oil ratio, will increase the availability of cracking for maximum conversion, assuming the unit is not already operating at a catalyst circulation limit. This can be achieved by increasing reactor heat load or switching to a lower coke selective (i.e., lower delta coke) catalyst. Reactor heat load can be raised by increased reactor temperature or lower feed preheat temperature. This, in turn, increases the catalyst–oil ratio to maintain the unit in heat balance.

The lower the *carbon on regenerated catalyst*, the higher the availability of cracking sites since less coke is blocking acid cracking sites. The carbon on the regenerated catalyst is reduced by increasing regeneration efficiency through the use of carbon monoxide oxidation promoters. Carbon on the regenerated catalyst can also be reduced by more efficient air and spent catalyst contact. Increased regenerator bed levels also reduce the amount of carbon on the regenerated catalyst through increased residence time but this must be traded off with reduced dilute phase disengager residence time and the possibility for increased catalyst losses.

3.4.2 Catalyst Treatment

The latest technique developed by the refining industry to increase gasoline yield and quality is to treat the catalysts from the cracking units to remove metal poisons that accumulate on the catalyst (Gerber et al., 1999). Nickel, vanadium, iron, and copper compounds contained in catalytic cracking feedstocks are deposited on the catalyst during the cracking operation, thereby adversely affecting both catalyst activity and selectivity. Increased catalyst metal contents affect catalytic cracking yields by increasing coke formation, decreasing gasoline and butane and butylene production, and increasing hydrogen production.

The recent commercial development and adoption of cracking catalyst-treating processes definitely improve the overall catalytic cracking process economics.

3.4.2.1 Demet Process

A cracking catalyst is subjected to two pretreatment steps. The first step affects vanadium removal; the second, nickel removal, to prepare the metals on the catalyst for chemical conversion to compounds (chemical treatment step) that can be readily removed through water washing (catalyst wash step). The treatment steps include use of a sulfurous compound followed by chlorination with an anhydrous chlorinating agent (e.g., chlorine gas) and washing with an aqueous solution of a chelating agent (e.g., citric acid, $HO_2CCH_2C(OH)$ $(CO_2H)CH_2CO_2H$, 2-hydroxy-1,2,3-propanetricarboxylic acid). The catalyst is then dried and further treated before returning to the cracking unit.

3.4.2.2 Met-X Process

This process consists of cooling, mixing, and ion-exchange separation, filtration, and resin regeneration. Moist catalyst from the filter is dispersed in oil and returned to the cracking reactor in a slurry. On a continuous basis, the catalyst from a cracking unit is cooled and then transported to a stirred reactor and mixed with an ion-exchange resin (introduced as slurry). The catalyst—resin slurry then flows to an elutriator for separation. The catalyst slurry is taken overhead to a filter, and the wet filter cake is slurried with oil and pumped into the catalytic cracked feed system. The resin leaves the bottom of the elutriator and is regenerated before returning to the reactor.

3.5 PROCESS OPTIONS

The processes described later are the evolutionary offspring of the FCC and the residuum catalytic cracking processes. Some of these newer processes use catalysts with different silica/alumina ratios as acid support of metals such as Mo, Co, Ni, and W. In general, the first catalyst used to remove metals from oils was the conventional HDS catalyst. Diverse natural minerals are also used as raw material for elaborating catalysts addressed to the upgrading of heavy fractions. Among these minerals are: clays; manganese nodules; bauxite activated with vanadium (V), nickel (Ni), chromium (Cr), iron (Fe), and cobalt (Co), as well as iron laterites and sepiolites; and mineral nickel and transition metal sulfides supported on silica and alumina. Other kinds of catalysts, such as vanadium sulfide, are generated *in situ*, possibly in colloidal states.

In the past decades, in spite of the difficulty of handling heavy feedstocks, residuum fluidized catalytic cracking (RFCC) has evolved to become a well-established approach for converting a significant portion of the heavier fractions of the crude barrel into a high-octane gasoline blending component. RFCC, which is an extension of conventional FCC technology for applications involving the conversion of highly contaminated residua, has been commercially proven on feedstocks ranging from gas oil residuum blends to atmospheric residua, as well as blends of atmospheric and vacuum residua blends. In addition to high gasoline yields, the RFCC unit also produces gaseous, distillate, and fuel oil range products.

The product quality from the residuum fluidized catalytic cracker is directly affected by its feedstock quality. In particular, and unlike hydrotreating, the RFCC redistributes sulfur among the various products but does not remove sulfur from the products unless, of course, one discounts the sulfur that is retained by any coke formed on the catalyst. Consequently, tightening product specifications have forced refiners to hydrotreat some, or all, of the products from the resid catalytic cracking unit. Similarly, in the future, the emissions of sulfur oxides (SO_x) from a resid catalytic cracker may become more of an obstacle for residue conversion projects. For these reasons, a point can be reached where the economic operability of the unit can be sufficient to justify hydrotreating the feedstock to the catalytic cracker.

As an integrated conversion block, residue hydrotreating and RFCC complement each other and can offset many of the inherent deficiencies related to residue conversion.

3.5.1 Aquaconversion Process

The aquaconversion process (Figure 3.3) is a hydrovisbreaking technology that uses catalyst-activated transfer of hydrogen from water added to the feedstock. Reactions that lead to coke formation are suppressed, and there is no separation of asphaltene-type material (Marzin et al., 1998). The important aspect of the aquaconversion technology is that it neither produces any solid by-product, such as coke, nor requires any hydrogen source or high-pressure equipment. In addition, the aquaconversion process can be implanted in the production area, and thus the need for external diluent and its transport over large distances is eliminated. Light distillates from the raw crude can be used as diluent for both the production and the desalting processes.

In addition some catalysts processes have been developed such as catalytic aquathermolysis which is used widely for upgrading heavy oil (Chen et al., 2009; Fan et al., 2004; Li et al., 2007; Wen et al., 2007). In this process, to maximize the upgrading effect, the suitable catalysts should be chosen.

3.5.2 Asphalt Residual Treating Process

The asphalt residual treating (ART) process is a process for increasing the production of transportation fuels and reduces heavy fuel oil production, without hydrocracking (Table 3.3; Bartholic, 1989).

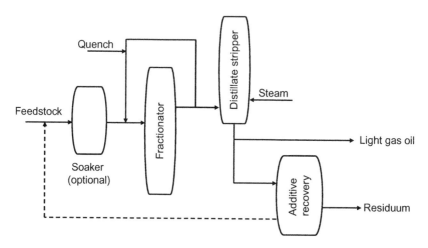

Figure 3.3 Aquaconversion process.

Table 3.3 Feedstock and Product Data for the ART Process

Feedstock	Mixed Residua[a]
API	14.9
Sulfur, wt%	4.1
Nitrogen, wt%	0.3
C5-asphaltenes	12.4
Carbon residue, wt%	15.8
Nickel, ppm	52.0
Vanadium, ppm	264.0
Products	
Naphtha, vol.%	62.1
No. 2 fuel oil, vol.%	35.5
No. 6 fuel oil, vol.%	2.3

[a]Blend of Arabian light vacuum residuum plus Arabian heavy vacuum residuum.

In this process, the preheated feedstock (which may be whole crude, atmospheric residuum, vacuum residuum, or tar sand bitumen) is injected into a stream of fluidized, hot catalyst (trade name: ArtCat). Complete mixing of the feedstock with the catalyst is achieved in the contactor, which is operated within a pressure–temperature envelope to ensure selective vaporization. The vapor and the contactor effluent are quickly and efficiently separated from each other and entrained hydrocarbons are stripped from the contaminant (containing spent solid) in the stripping section. The contactor vapor effluent and vapor from the stripping section are combined and rapidly quenched in a quench drum to minimize product degradation. The cooled products are then transported to a conventional fractionator that is similar to that found in an FCC unit. Spent solid from the stripping section is transported to the combustor bottom zone for carbon burn off.

Contact of the heavy feedstock with the fluidizable catalyst in a short residence time contactor causes the lower boiling components of the feedstock to vaporize and asphaltene constituents (high-molecular-weight compounds) are cracked to yield lower boiling compounds and coke. The metals present, as well as some of the sulfur and the nitrogen compounds in the nonvolatile constituents, are retained on the catalyst. At the exit of the contacting zone, the oil vapors are separated from the catalyst and are rapidly quenched to minimize thermal cracking of the products. The catalyst, which holds metals, sulfur, nitrogen, and coke, is transferred to the regenerator where the combustible portion is oxidized and removed. Regenerated contact material, bearing metals but very little coke, exits the regenerator and passes to the contactor for further removal of contaminants from the charge stock.

In the combustor, coke is burned from the spent solid that is then separated from combustion gas in the surge vessel. The surge vessel circulates regenerated catalyst streams to the contactor inlet for feed vaporization and to the combustor bottom zone for premixing.

The components of the combustion gases include carbon dioxide (CO_2), nitrogen (N_2), oxygen (O_2), sulfur oxides (SO_x), and nitrogen oxides (NO_x) that are released from the catalyst with the combustion of the coke in the combustor. The concentration of sulfur oxides in the combustion gas requires treatment for their removal.

3.5.3 Heavy Oil Treating Process

The heavy oil treating (HOT) process is a catalytic cracking process for upgrading heavy feedstocks such as topped crude oils, vacuum residua, and solvent deasphalted bottoms using a fluidized bed of iron ore particles (Table 3.4).

The main section of the process consists of three fluidized reactors and separate reactions take place in each reactor (*cracker, regenerator,* and *desulfurizer*):

$$Fe_3O_4 + \text{asphaltene constituents} \rightarrow \text{coke}/Fe_3O_4 + \text{oil} + \text{gas} \quad \text{(in the } cracker)$$
$$3FeO + H_2O \rightarrow Fe_3O_4 + H_2 \text{ (in the } cracker)$$
$$\text{coke}/Fe_3O_4 + O_2 \rightarrow 3FeO + CO + CO_2 \text{ (in the } regenerator)$$
$$FeO + SO_2 + 3CO \rightarrow FeS + 3CO_2 \text{ (in the } regenerator)$$
$$3FeS + 5O_2 \rightarrow Fe_3O_4 + 3SO_2 \text{ (in the } desulfurizer)$$

In the *cracker*, heavy oil cracking and the steam−iron reaction take place simultaneously under the conditions similar to usual thermal cracking. Any unconverted feedstock is recycled to the cracker from the bottom of the scrubber. The scrubber effluent is separated into

Table 3.4 Feedstock and Product Data for the HOT Process	
Feedstock	Arabian Light Vacuum Residuum
API	7.1
Sulfur, wt%	4.2
Carbon residue, wt%	21.6
Products	
Light naphtha (C5—180°C, C5—355°F), vol.%	15.2
Heavy naphtha (180–230°C, 355–445°F), vol.%	8.2
Light gas oil (230–360°C, 445–680°F), vol.%	2.3
Heavy gas oil (360–510°C, 680–950°F), vol.%	28.2

hydrogen gas, LPG, and liquid products that can be upgraded by conventional technologies to priority products.

In the *regenerator*, coke deposited on the catalyst is partially burned to form carbon monoxide in order to reduce iron tetroxide and to act as a heat supply. In the *desulfurizer*, sulfur in the solid catalyst is removed and recovered as molten sulfur in the final recovery stage.

3.5.4 R2R Process

The R2R process is an FCC process for conversion of heavy feedstocks (Heinrich and Mauleon, 1994; Inai, 1994). In this process, the feedstock is vaporized upon contacting hot regenerated catalyst at the base of the riser and lifts the catalyst into the reactor vessel separation chamber where rapid disengagement of the hydrocarbon vapors from the catalyst is accomplished by both a special solids separator and cyclones. The bulk of the cracking reactions takes place at the moment of contact and continues as the catalyst and hydrocarbons travel up the riser. The reaction products, along with a minute amount of entrained catalyst, then flow to the fractionation column. The stripped spent catalyst, deactivated with coke, flows into the Number 1 regenerator.

Partially regenerated catalyst is pneumatically transferred via an air riser to the Number 2 regenerator, where the remaining carbon is completely burned in a dryer atmosphere. This regenerator is designed to minimize catalyst inventory and residence time at high temperature while optimizing the coke-burning rate. Flue gases pass through external cyclones to a waste heat recovery system. Regenerated catalyst flows into a withdrawal well and after stabilization is charged back to the oil riser.

3.5.5 Reduced Crude Oil Conversion Process

The RCC process arose because of a trend for low-boiling products; most refineries perform the operation by partially blending residua into vacuum gas oil. However, conventional FCC processes have limits in residue processing, so residue FCC processes have lately been employed one after another. Because the residue FCC process enables efficient gasoline production directly from residues, it will play the most important role as a residue cracking process, along with the residue hydroconversion process. Another role of the *residuum FCC process* is to generate high-quality gasoline blending stock and petrochemical feedstock. Olefins (propene, butenes, and pentenes) serve as feed for alkylating processes, for polymer gasoline, as well as for additives for reformulated gasoline.

In the RCC process (Table 3.5), the clean regenerated catalyst enters the bottom of the reactor riser where it contacts low-boiling hydrocarbon *lift gas* that accelerates the catalyst up the riser prior to feed injection. At the top of the lift gas zone, the feed is injected through a series of nozzles located around the circumference of the reactor riser.

Table 3.5 Feedstock and Product Data for the RCC Process			
Feedstock	Residuum[a]	Residuum[a]	Residuum[a]
API	22.8	21.3	19.2
Sulfur, wt%			
Nitrogen, wt%			
Carbon residue, wt%	0.2	6.4	5.5
Nickel, ppm	1.0	22.0	34.0
Vanadium, ppm			
Products			
Gasoline (C5—221°C, C5—430°F), vol.%	59.1	56.6	55.6
Light cycle oil (221–322°C, 430–610°F), vol.%	16.3	15.4	16.3
Gas oil (> 322°C, >610°F), vol.%	6.2	9.0	9.6
Coke, wt%	8.4	9.1	10.8
[a]Unspecified.			

The catalyst—oil disengaging system is designed to separate the catalyst from the reaction products and then rapidly remove the reaction products from the reactor vessel. Spent catalyst from reaction zone is first steam stripped, to remove adsorbed hydrocarbon, and then routed to the regenerator. In the regenerator, all of the carbonaceous deposits are removed from the catalyst by combustion and restoring the catalyst to an active state with a very low carbon content. The catalyst is then returned to the bottom of the reactor riser at a controlled rate to achieve the desired conversion and selectivity to the primary products.

3.5.6 Residue FCC Process

This process is a version of the FCC process that has been adapted to conversion of residua that contain high amounts of metal and asphaltene constituents. (Speight and Ozum, 2002).

In this process, a residuum is desulfurized and the nonvolatile fraction from the HDS unit is charged to the residuum FCC unit. The reaction system is an external vertical riser terminating in a closed cyclone system. Dispersion steam in amounts higher than that used for gas oils is used to assist in the vaporization of any volatile constituents of heavy feedstocks.

The reaction system is an external vertical riser providing very low contact times and terminating in the riser cyclones for rapid separation of catalyst and vapors. A two-stage stripper is utilized to remove hydrocarbons from the catalyst. Hot catalyst flows at low velocity in dense phase through the catalyst cooler and returns to the regenerator. Regenerated catalyst flows to the bottom of the riser to meet the feed. A two-stage stripper is utilized to remove hydrocarbons from the catalyst. Hot catalyst flows at low velocity in dense phase through the catalyst cooler and returns to the regenerator. Regenerated catalyst flows to the bottom of the riser to meet the feed.

The coke deposited on the catalyst is burned off in the regenerator along with the coke formed during the cracking of the gas oil fraction. If the feedstock contains high proportions of metals, control of the metals on the catalyst requires excessive amounts of catalyst withdrawal and fresh catalyst addition. This problem can be addressed by feedstock pretreatment. Regenerator bed temperatures are limited to around 730°C (1300°F) and feed introduction systems are designed for efficient mixing of oil and catalyst and rapid quenching of catalyst temperature to the equilibrium mix temperature.

The feedstocks for the process are rated on the basis of carbon residue and content of metals. Thus, *good quality feedstocks* have less than 5% by weight carbon residue and less than 10 ppm metals. *Medium quality feedstocks* have greater than 5% but less than 10% by weight carbon residue and greater than 10% by weight but less than 30 ppm metals. *Poor quality feedstocks* have greater than 10% but less than 20% by weight carbon residue and greater than 30% by weight but less than 150 ppm metals. Finally, *bad quality feedstocks* have greater than 20% by weight carbon residue and greater than 150 ppm metals. One might question the value of this rating of the feedstocks for the HOC process since these feedstock ratings can apply to many FCC processes.

3.5.7 Shell FCC Process
The Shell FCC process is designed to maximize the production of distillates from residua and other heavy feedstocks (Table 3.6). In this process, the preheated feedstock (vacuum gas oil, atmospheric residuum) is mixed with the hot regenerated catalyst. After reaction in a riser, volatile materials and catalyst are separated after which the spent catalyst is immediately stripped of entrained and adsorbed hydrocarbons in a very effective multistage stripper. The stripped catalyst gravitates through a short standpipe into a single vessel, a simple, reliable, and yet efficient catalyst regenerator. Regenerative flue gas passes via a cyclone/swirl tube combination to a power recovery turbine. From

Table 3.6 Feedstock and Product Data for the Shell FCC Process		
Feedstock	Residuum[a]	Residuum[a]
API	18.2	13.4
Sulfur, wt%	1.1	1.3
Carbon residue, wt%	1.2	4.7
Products		
Gasoline (C5—221°C, C5—430°F), wt%	49.5	46.2
Light cycle oil (221—370°C, 430—700°F), wt%	20.1	19.1
Heavy cycle oil (>370°C, >700°F), wt%	5.9	10.8
Coke, wt%	5.9	7.6
[a]Unspecified		

the expander turbine, the heat in the flue gas is further recovered in a waste heat boiler. Depending on the environmental conservation requirements, a *deNO$_x$ing, deSO$_x$ing*, and particulate emission control device can be included in the flue gas train.

There is a claim that feedstock pretreatment of bitumen (by hydrogenation) prior to FCC (or for that matter any catalytic cracking process) can result in enhanced yield of naphtha. It is suggested that mild hydrotreating be carried out upstream of an FCC unit to provide an increase in yield and quality of distillate products. This is in keeping with observations that mild hydrotreating of bitumen was reported to produce low-sulfur liquids that would be amenable to further catalytic processing (Figure 3.4; Speight, 2007, 2011; Speight and Ozum, 2002).

3.5.8 S&W Fluid Catalytic Cracking Process

The *S&W FCC process* is also designed to maximize the production of distillates from residua (Table 3.7). In this process, the heavy feedstock is injected into a stabilized, upward flowing catalyst stream whereupon the feedstock–steam–catalyst mixture travels up the riser and is separated by a high efficiency inertial separator. The product vapor goes overhead to the main fractionator.

The spent catalyst is immediately stripped in a staged, baffled stripper to minimize hydrocarbon carryover to the regenerator system. The first regenerator (650–700°C, 1200–1290°F) burns 50–70% of the coke in an incomplete carbon monoxide combustion mode running

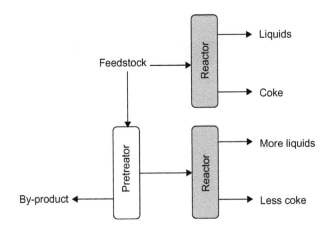

Figure 3.4 Potential pretreating scheme for catalytic cracker feedstocks.

Table 3.7 Feedstock and Product Data for the S&W FCC Process

Feedstock	Residuum[a]	Residuum[a]
API	24.1	22.3
Sulfur, wt%	0.8	1.0
Carbon residue, wt%	4.4	6.5
Products		
Naphtha, vol.%	61.5	60.2
Light cycle oil, vol.%	16.6	17.5
Heavy cycle oil, vol.%	5.6	6.6
Coke, wt%	7.1	7.8
Conversion, vol.%	77.7	75.9

[a]Unspecified.

countercurrently. This relatively mild, partial regeneration step mini-mizes the significant contribution of hydrothermal catalyst deactiva-tion. The remaining coke is burned in the second regenerator (*ca.* 775°C, 1425°F) with an extremely low steam content. Hot clean cata-lyst enters a withdrawal well that stabilizes its fluid qualities prior to being returned to the reaction system.

3.5.9 Microwave Processes

Finally, the use of microwaves to generate heat in a uniform and con-trolled fashion is well known and microwave technology with catalysts is another technology which has been applied recently in upgrading of heavy oil (Lam et al., 2012; Mutyala et al., 2010). For example, catalytic hydroconversion of residue from coal liquefaction by using microwave irradiation with a Ni catalyst (Wang et al., 2008), and microwave-assisted desulfurization of heavy and sour crude oil using iron powder as catalyst (Leadbeater and Khan, 2008). Also upgrading of Athabasca bitumen with microwave-assisted catalytic pyrolysis was carried out in one study. Silicon carbide is used for pyrolysis and nickel and molybdenum nanoparticles as catalysts used to enhance the pyrol-ysis performance. The results of the work suggest that microwave heat-ing with nanoparticles catalyst can be a useful tool for the upgrading of heavy crudes such as bitumen because of rapid heating and energy efficient (Jeon et al., 2012).

REFERENCES

Avidan, A.A., Krambeck, F.J., 1990. FCC closed cyclone system eliminates post riser cracking. Proceedings of the Annual Meeting on National Petrochemical and Refiners Association.

Bartholic, D.B., 1981a. Preparation of FCC Charge from Residual Fractions. United States Patent 4,243,514. January 6.

Bartholic, D.B., 1981b. Upgrading Petroleum and Residual Fractions Thereof. United States Patent 4,263,128. April 21.

Bartholic, D.B., 1989. Process for Upgrading Tar Sand Bitumen. United States Patent 4,804,459. February 14.

Chen, Y.L., Wang, Y.Q., Lu, J.Y., Wu, C., 2009. The viscosity reduction in catalytic aquathermolysis of heavy oil. Fuel 88, 1426–1434.

Fan, H., Zhang, Y., Lin, Y., 2004. The catalytic effects of minerals on aquathermolysis of heavy oils. Fuel 83, 2035–2039.

Gary, J.H., Handwerk, G.E., Kaiser, M.J., 2007. Petroleum Refining: Technology and Economics, fifth ed. CRC Press, Taylor & Francis Group, Boca Raton, FL.

Gerber, M.A., Fulton, J.L., Frye, J.G., Silva, L.J., Bowman, L.E., Wai, C.M., 1999. Regeneration of Hydrotreating and FCC Catalysts. Report No. PNNL-13025. US Department of Energy Contract No. DE-AC06-76RLO 1830. Pacific Northwest National Laboratory, Richland, Washington.

Heinrich, G., Mauleon, J.-L., 1994. The R2R process: 21st century FCC technology. Rev. Inst. Fr. du Pétrole 49 (5), 509–520.

Hsu, C.S., Robinson, P.R. (Eds.), 2006. Practical Advances in Petroleum Processing, vols 1 and 2. Springer Science, New York, NY.

Inai, K., 1994. Operation results of the R2R process. Rev. Inst. Fr. du Pétrole 49 (5), 521–527.

Jeon, S.G., Kwak, N.S., Rho, N.S., Ko, C.H., Na, J.-G., Yi, K.B., et al., 2012. Catalytic pyrolysis of Athabasca bitumen in H_2 atmosphere using microwave irradiation. Chem. Eng. Res. Des. 90, 1292–1296.

Lam, S.S., Russell, A.D., Lee, C.L., Chase, H.A., 2012. Microwave-heated pyrolysis of waste automotive engine oil: influence of operation parameters on the yield, composition, and fuel properties of pyrolysis oil. Fuel 92, 327–339.

Leadbeater, N.E., Khan, M.R., 2008. Microwave-promoted desulfurization of heavy oil and a review of recent advances on process technologies for upgrading of heavy and sulfur containing crude oil. Energy Fuels 22, 1836–1839.

Li, W., Zhu, J.H., Qi, J.H., 2007. Application of nano-nickel catalyst in the viscosity reduction of Liaohe extra-heavy oil by aquathermolysis. J. Fuel Chem. Technol. 35, 176–180.

Long, S.L., Johnson, A.R., Dharia, D., 1993. Advances in Residual Oil FCC. of the Annual Meeting on National Petrochemical and Refiners Association. Paper AM-93-50.

Marzin, R., Pereira, P., Zacarias, L., Rivas, L. McGrath, M., Thompson, G.J 1998. Resid Conversion through the Aquaconversion Technology – An Economical and Environmental Solution. SPE Paper No. 1998.086.

McPherson, L.J., 1984. Causes of FCC reactor coke deposits identified. Oil Gas J. September 10, 139.

Mutyala, S., Fairbridge, C., Jocelyn Paré, J.R., Bélanger, J.M.R., Ng, S., Hawkins, R., 2010. Microwave applications to oil sands and petroleum: a review. Fuel Process. Technol. 91, 127–135.

Speight, J.G., 2000. The Desulfurization of Heavy Oils and Residua, second ed. Marcel Dekker, New York, NY.

Speight, J.G., 2004. New approaches to hydroprocessing. Catal. Today 98 (1−2), 55−60.

Speight, J.G., 2007. The Chemistry and Technology of Petroleum, fourth ed. CRC Press, Taylor & Francis Group, Boca Raton, FL.

Speight, J.G., 2011. The Refinery of the Future. Gulf Professional Publishing, Elsevier, Oxford, UK.

Speight, J.G., Ozum, B., 2002. Petroleum Refining Processes. Marcel Dekker, New York, NY.

Wang, T.X., Zong, Z.M., Zhang, V.W., Wei, Y.B., Zhao, W., Li, B.M., et al., 2008. Microwave-assisted hydroconversion of demineralized coal liquefaction residues from Shenfu and Shengli coals. Fuel 87, 498−507.

Wen, S.B., Zhao, Y.J., Liu, Y.J., Hu, S.B., 2007. A study on catalytic aquathermolysis of heavy crude oil during steam stimulation. In: Proceedings of the International Symposium on Oilfield Chemistry. Paper No. 106180-MS, 2007, February 28−March 2, Houston, TX.

CHAPTER 4

Hydrotreating and Desulfurization

4.1 INTRODUCTION

Catalytic hydrotreating, also referred to as *hydroprocessing* or *hydrodesulfurization* (*HDS*), commonly appears in multiple locations in a refinery (Dolbear, 1998; Gary et al., 2007; Hsu and Robinson, 2006; Meyers, 1997; Speight, 2000, 2007; Speight and Ozum, 2002). Furthermore, *hydrotreating*, a term often used synonymously with HDS, is a catalytic refining process widely used to remove sulfur from petroleum products such as naphtha, gasoline, diesel fuel, kerosene, and fuel oil (Gary et al., 2007; Hsu and Robinson, 2006; Meyers, 1997; Speight, 2000, 2007; Speight and Ozum, 2002). The objective of the hydrotreating process is to remove sulfur as well as other unwanted compounds, for example, unsaturated hydrocarbons and nitrogen from refinery process streams. HDS processes typically include facilities for the capture and removal of the resulting hydrogen sulfide (H_2S) gas, which is subsequently converted into by-product elemental sulfur or sulfuric acid.

A growing dependence on high-heteroatom heavy oils and heavy feedstocks has emerged as a result of continuing decreasing availability of conventional crude oil through the worldwide depletion of petroleum reserves. Thus, the ever-growing tendency to convert as much as possible of lower grade feedstocks to liquid products is causing an increase in the total sulfur content in penultimate products. Refiners must, therefore, continue to remove substantial portions of sulfur from the lighter products so that the final products meet specifications. However, heavy feedstocks, heavy crude oil, and tar sand bitumen pose a particularly difficult problem in terms of the propensity of these feedstocks to form coke and to shorten catalyst life (Ancheyta and Speight, 2007; Speight, 2000).

Catalytic hydrotreating is also seeing increasing use prior to catalytic cracking to reduce sulfur and improve product yields, and to upgrade middle-distillate petroleum fractions into finished kerosene, diesel fuel, and heating fuel oils. In addition, hydrotreating converts olefins and aromatics to saturated compounds. At one time, hydrotreating a heavy feedstock was not even considered because of the hydrogen demands and the detrimental effect of the feedstock on the catalyst.

Finally, hydrotreating of heavy feedstocks requires considerably different catalysts and process flows, depending on the specific operation so that efficient hydroconversion through uniform distribution of liquid, hydrogen-rich gas, and catalyst across the reactor is assured. There will also be automated demetallization of fixed-bed systems as well as more units that operate as ebullating-bed hydrocrackers (Chapter 5).

4.2 PROCESS TYPES

Hydrotreating is carried out by charging the feed to the reactor with hydrogen and suitable catalysts, such as tungsten−nickel sulfide, cobalt−molybdenum−alumina, nickel oxide−silica−alumina, and platinum−alumina. Most processes employ cobalt−molybdenum catalysts, which generally contain about 10% by weight molybdenum oxide and less than 1% by weight cobalt oxide supported on alumina. The temperatures employed are in the range of 300−345°C (570−850°F), and the hydrogen pressures are about 500−1000 psi.

Although many different hydrotreater designs are marketed, they all work along the same principle—all processes use the reaction of hydrogen with the hydrocarbon feedstock to produce hydrogen sulfide and a desulfurized hydrocarbon product (Gary et al., 2007; Hsu and Robinson, 2006; Speight, 2007). This also invokes the concept of different process types (usually names after the types of reactor employed). Each reactor has its own merits and, providing the choice made according to the feedstock has an excellent chance of producing the desired products.

The reaction temperature typically should be on the order of 290−445°C (550−850°F) with a hydrogen gas pressures on the order of 150 and 3,000 psi—the low temperature minimizes cracking reactions. In some designs, the feedstock is heated and then mixed with the hydrogen rather than the option of passing moderately heated feedstock (i.e., feedstock not at the full reactor temperature) and moderately heated hydrogen (i.e., hydrogen not at the full reactor temperature) into the reactor. The gas mixture is led over a catalyst bed of metal oxides (most often cobalt or molybdenum oxides on different metal carriers). The catalysts help the hydrogen to react with sulfur and nitrogen to form hydrogen sulfides (H_2S) and ammonia. The reactor effluent is then cooled, and the oil feed and gas mixture are then separated in a stripper column. Part of the stripped gas may be recycled to the reactor.

4.2.1 Downflow Fixed-Bed Reactor

The reactor design commonly used in HDS of distillates is the fixed-bed reactor design in which the feedstock enters at the top of the reactor and the product leaves at the bottom of the reactor. The catalyst remains in a stationary position (fixed bed) with hydrogen and petroleum feedstock passing in a downflow direction through the catalyst bed. The HDS reaction is exothermic and the temperature rises from the inlet to the outlet of each catalyst bed. With a high hydrogen consumption and subsequent large temperature rise, the reaction mixture can be quenched with cold recycled gas at intermediate points in the reactor system. This is achieved by dividing the catalyst charge into a series of catalyst beds and the effluent from each catalyst bed is quenched to the inlet temperature of the next catalyst bed.

The extent of desulfurization is controlled by raising the inlet temperature to each catalyst bed to maintain constant catalyst activity over the course of the process. Fixed-bed reactors are mathematically modeled as plug-flow reactors with very little back-mixing in the catalyst beds. The first catalyst bed is poisoned with vanadium and nickel at the inlet to the bed and may be a cheaper catalyst (*guard bed*). As the catalyst is poisoned in the front of the bed, the temperature exotherm moves down the bed and the activity of the entire catalyst charge declines thus requiring a rise in the reactor temperature over the course of the process sequence. After catalyst regeneration, the reactors are opened and inspected, and the high-metal-content catalyst layer at the inlet to the first bed may be discarded and replaced with fresh catalyst. The catalyst loses activity after a series of regenerations and, consequently, after a series of regenerations it is necessary to replace the complete catalyst charge. In the case of very-high-metal-content feedstocks (such as heavy feedstocks), it is often necessary to replace the entire catalyst charge rather than to regenerate it. This is due to the fact that the metal contaminants cannot be removed by economical means during rapid regeneration and the metals have been reported to interfere with the combustion of carbon and sulfur, catalyzing the conversion of sulfur dioxide (SO_2) to sulfate (SO_4^{2-}), which has a permanent poisoning effect on the catalyst.

Fixed-bed HDS units are generally used for distillate HDS and may also be used for heavy feedstock HDS but require special precautions in processing. The heavy feedstock must undergo two-stage electrostatic desalting so that salt deposits do not plug the inlet to the first catalyst bed and the heavy feedstock must be low in vanadium and nickel content to avoid plugging the beds with metal deposits. Hence the need for a guard bed in heavy feedstock HDS reactors.

During the operation of a fixed-bed reactor, contaminants entering with fresh feed are filtered out and fill the voids between catalyst particles in the bed. The buildup of contaminants in the bed can result in the channeling of reactants through the bed thereby reducing the HDS efficiency. As the flow pattern becomes distorted or restricted the pressure drop throughout the catalyst bed increases. If the pressure drop becomes high enough, physical damage to the internal parts of the reactor can occur. When high-pressure drops are observed throughout any portion of the reactor, the unit is shut down and the catalyst bed is skimmed and refilled.

With fixed-bed reactors, a balance must be reached between reaction rate and pressure drop across the catalyst bed. As catalyst particle size is decreased, the desulfurization reaction rate increases but so does the pressure drop across the catalyst bed. Expanded-bed reactors do not have this limitation and small 1/32 in. (0.8 mm) extrudate catalysts or fine catalysts may be used without increasing the pressure drop.

4.2.2 Upflow Expanded-Bed Reactor

Expanded-bed reactors are applicable to heavy feedstocks and are commercially used for very heavy, high metals, and/or dirty feedstocks having extraneous fine solid material. They operate in such a way that the catalyst is in an expanded state so that the extraneous solids pass through the catalyst bed without plugging. They are isothermal, which conveniently handles the high-temperature exotherms associated with high hydrogen consumptions. Since the catalyst is in an expanded state of motion, it is possible to treat the catalyst as a fluid and to withdraw and add catalyst during operation.

Expanded beds of catalyst are referred to as particulate fluidized beds insofar as the feedstock and hydrogen flow upward through an expanded bed of catalyst with each catalyst particle in independent motion. Thus, the catalyst migrates throughout the entire reactor bed. Spent catalyst may be withdrawn and replaced with fresh catalyst on a daily basis. Daily catalyst addition and withdrawal eliminate the need for costly shutdowns to change out catalyst and also result in a constant equilibrium catalyst activity and product quality. The catalyst is withdrawn daily and has vanadium, nickel, and carbon content that is representative on a macroscale of what is found throughout the entire reactor. On a microscale, individual catalyst particles have ages from that of fresh catalyst to as old as the initial catalyst charge to the unit but the catalyst particles of each age group are so well dispersed in the reactor that the reactor contents appear uniform.

In the unit, the feedstock and hydrogen recycle gas enter the bottom of the reactor, pass up through the expanded catalyst bed, and leave from the top of the reactor. Commercial expanded-bed reactors normally operate with 1/32 in. (0.8 mm) extrudate catalysts that provide a higher rate of desulfurization than the larger catalyst particles used in fixed-bed reactors. With extrudate catalysts of this size, the upward liquid velocity based on fresh feedstock is not sufficient to keep the catalyst particles in an expanded state. Therefore, for each part of the fresh feed, several parts of product oil are taken from the top of the reactor, recycled internally through a large vertical pipe to the bottom of the reactor, and pumped back up through the expanded catalyst bed. The amount of catalyst bed expansion is controlled by the recycle of product oil back up through the catalyst bed.

The expansion and turbulence of gas and oil passing upward through the expanded catalyst bed are sufficient to cause almost complete random motion in the bed (particulate fluidized). This effect produces the isothermal operation. It also causes almost complete backmixing. Consequently, in order to effect near complete sulfur removal (over 75%), it is necessary to operate with two or more reactors in series. The ability to operate at a single temperature throughout the reactor or reactors, and to operate at a selected optimum temperature rather than an increasing temperature from the start to the end of the run, results in more effective use of the reactor and catalyst contents. When all these factors are put together, that is, use of a smaller catalyst particle size, isothermal, fixed temperature throughout run, back-mixing, daily catalyst addition, and constant product quality, the reactor size required for an expanded bed is often smaller than that required for a fixed bed to achieve the same product goals. This is generally true when the feeds have high initial boiling points and/or the hydrogen consumption is very high.

4.2.3 Demetallization Reactor

Whilst not necessarily a process in its own right, the demetallization process (guard-bed process and guard-bed reactor) is typically used for heavy feedstocks that have a relatively high metals contents (>300 ppm) which substantially increases catalyst consumption because the metals poison the catalyst, thereby requiring frequent catalyst replacement.

The usual desulfurization catalysts are relatively expensive for these consumption rates but there are catalysts that are relatively inexpensive and can be used in the first reactor to remove a large percentage of the metals. Subsequent reactors downstream of the first reactor would use normal HDS catalysts. Since the catalyst materials are proprietary, it is not possible to identify them here. However, it is understood that such catalysts contain little or no metal promoters, that is, nickel, cobalt, and molybdenum. Metals removal on the order of 90% has been observed with these materials.

Thus, one method of controlling demetallization is to employ separate smaller *guard reactors* just ahead of the fixed-bed HDS reactor section. The preheated feed and hydrogen pass through the guard reactors that are filled with an appropriate catalyst for demetallization that is often the same as the catalyst used in the HDS section. The advantage of this system is that it enables replacement of the most contaminated catalyst (*guard bed*), where pressure drop is highest, without having to replace the entire inventory or shut down the unit. The feedstock is alternated between guard reactors while catalyst in the idle guard reactor is being replaced.

When the expanded-bed design is used, the first reactor could employ a low-cost catalyst (5% of the cost of Co/Mo catalyst) to remove the metals and subsequent reactors can use the more selective HDS catalyst. The demetallization catalyst can be added continuously without taking the reactor out of service, and the spent demetallization catalyst can be loaded to more than 30% vanadium, which makes it a valuable source of vanadium.

4.3 PROCESS VARIABLES

In the process, the feedstock is preheated and mixed with hot recycle gas containing hydrogen and the mixture is passed over the catalyst in the reactor section at temperatures between 290°C and 445°C (550–850°F) and pressures between 150 and 3,000 psi (Table 4.1). The reactor effluent is then cooled by heat exchange, and desulfurized liquid hydrocarbon product and recycle gas are separated at essentially the same pressure as used in the reactor. The recycle gas is then scrubbed and/or purged of the hydrogen sulfide and light hydrocarbon gases, mixed with fresh hydrogen makeup, and preheated prior to mixing with hot hydrocarbon feedstock.

Although hydrocracking will occur during hydrotreating, attempts are made to minimize such effects but the degree of cracking is dependent on the nature of the feedstock. These are (i) process flow, (ii) feedstock properties, (iii) reaction temperature, (iv) hydrogen particle pressure, (v) gas recycle, and (vi) coke formation.

4.3.1 Process Flow

The choice of processing schemes for a given hydrotreating application depends upon the nature of the feedstock as well as the product requirements (Gary et al., 2007; Hsu and Robinson, 2006; Speight, 2007, 2011; Speight and Ozum, 2002). For the heavy feedstocks, the process is usually hydrotreating followed by catalytic cracking or by

Table 4.1 Process Parameters for HDS	
Parameter	**Heavy Feedstocks**
Temperature (°C)	340–425
Pressure (atm.)	55–170
Liquid hourly space velocity (LHSV)	0.2–1.0
H_2 recycle rate (ft^3/bbl)	3000–5000
Catalysts life (years)	0.5–1.0
Sulfur removal (%)	85.0
Nitrogen removal (%)	40.0
Source: Speight (2007).	

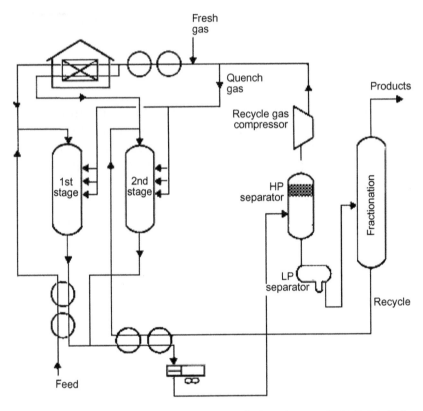

Figure 4.1 A single-stage or two-stage (optional) hydrotreating/hydrocracking unit. OSHA Technical Manual, Section IV, Chapter 2: Petroleum Refining Processes. http://www.osha.gov/dts/osta/otm/otm_iv/otm_iv_2.html.

hydrocracking, and the hydrotreating segment can (as with hydro-cracking) be illustrated as a single-stage or as a two-stage operation (Figure 4.1). Variations to the process are feedstocks dependent.

For example, the single-stage process can be used to produce gaso-line but is more often used to produce middle distillate from heavy vacuum gas oils. The two-stage process was developed primarily to produce high yields of gasoline from straight-run gas oil, and the first stage may actually be a purification step to remove sulfur-containing (as well as nitrogen-containing) organic materials. Both processes use an extinction-recycling technique to maximize the yields of the desired product. Significant conversion of heavy feedstocks can be accom-plished by hydrocracking at high severity (Howell et al., 1985). For some applications, the products boiling up to 340°C (650°F) can be blended to give the desired final product.

4.3.2 Feedstock Properties

The character of the *feedstock properties*, especially the feed boiling range, has a definite effect on the ultimate design of the desulfurization unit and process flow. High-boiling feedstocks such as heavy oil, extra-heavy oil, and tar sand bitumen are not usually hydrotreating feedstocks but the increased amount of heavy feedstocks in refineries has caused refiners to seek methods of rendering these feedstocks suitable as feedstocks for catalytic cracking units. And, as with lower boiling feedstocks, there is a definite relationship between the percent by weight sulfur in the feedstock and the hydrogen requirements.

One of the issues that arises when hydroprocessing heavy feedstocks is the deposition of metals, in particular vanadium, on the catalyst. It is not possible to remove vanadium from the catalyst, which must therefore be replaced when deactivated, and the time taken for catalyst replacement can significantly reduce the unit time efficiency. Fixed-bed catalysts tend to plug owing to solids in the feed or carbon deposits when processing heavy feedstocks. In addition, the exothermic reaction at high conversion gives difficult reactor design problems in heat removal and temperature control.

The problems encountered in hydrotreating heavy feedstocks can be directly equated to the amount of complex, higher boiling constituents that may require pretreatment (Reynolds and Beret, 1989; Speight, 2007; Speight and Moschopedis, 1979; Speight and Ozum, 2002). Processing these feedstocks is not merely a matter of applying know-how derived from refining *conventional* crude oils but requires knowledge of the composition (Ancheyta and Speight, 2007; Speight, 2007). The materials are not only complex in terms of the carbon number and boiling point ranges but also because a large part of this *envelope* falls into a range of model compounds and very little is known about the properties. It is also established that the majority of the higher molecular weight materials produce coke (with some liquids) but the majority of the lower molecular weight constituents produce liquids (with some coke).

It is the physical and chemical composition of a feedstock that plays a large part not only in determining the nature of the products that arise from refining operations but also in determining the precise manner by which a particular feedstock should be processed (Speight, 2007; Speight and Ozum, 2002). Furthermore, it is apparent that the conversion of heavy feedstocks requires new lines of thought to develop suitable processing scenarios. Indeed, the use of thermal (*carbon rejection*) processes and of hydrothermal (*hydrogen addition*) processes, which were inherent in the refineries designed to process lighter feedstocks, has been a particular cause for concern. This has brought about the evolution of processing schemes that accommodate the heavier feedstocks (Boening et al., 1987; Khan, 1998).

Catalyst life depends on the charge stock properties and the degree of desulfurization desired. The only permanent poisons to the catalyst are metals in the feedstock that deposit on the catalyst, usually quantitatively, causing permanent deactivation as they accumulate. However, this is usually of some concern when heavy feedstocks are used since they contain high amounts of metals.

In summary, hydrotreating processes differ depending upon the feedstock available and catalysts used, even within the subcategories of heavy feedstocks as used in this text (Chapter 1).

4.3.3 Reaction Temperature
A higher *reaction temperature* increases the rate of desulfurization at constant feed rate and the start-of-run temperature is set by the design desulfurization level, space velocity, and hydrogen partial pressure. The capability to increase temperature as the catalyst deactivates is built into most process or unit designs. Temperatures of 415°C (780°F) and above result in excessive coking reactions and higher than normal catalyst aging rates. Therefore, units are designed to avoid the use of such temperatures for any significant part of the cycle's life.

As a general rule, excessively high reactor temperature (and contact time) temperature will stimulate coke formation.

4.3.4 Hydrogen Partial Pressure
The important effect of *hydrogen partial pressure* is the minimization of coking reactions. If the hydrogen pressure is too low for the required duty at any position within the reaction system, premature aging of the remaining portion of catalyst will be encountered. In addition, the effect of hydrogen pressure on desulfurization varies with feed boiling range. For a given feed there exists a threshold level above which hydrogen pressure is beneficial to the desired desulfurization

reaction. Below this level, desulfurization drops off rapidly as hydrogen pressure is reduced.

One of the issue-related problems in processing of high-sulfur and high-nitrogen feedstocks is the large quantity of hydrogen sulfide (H_2S) and ammonia (NH_3) that are produced. Substantial removal of both compounds from the recycle gas can be achieved by the injection of water in which, under the high-pressure conditions employed, both hydrogen sulfide and ammonia are very soluble compared with hydrogen and hydrocarbon gases. The solution is processed in a separate unit for the recovery of anhydrous ammonia and hydrogen sulfide.

As the *space velocity* is increased, desulfurization is decreased but increasing the hydrogen partial pressure and/or the reactor temperature can offset the detrimental effect of increasing space velocity.

Hydrogen consumption is also a parameter that varies with feedstock composition (Ancheyta and Speight, 2007; Speight, 2007; Speight and Ozum, 2002), thereby indicating the need for a thorough understanding of the feedstock constituents if the process is to be employed to maximum efficiency. A convenient means of understanding the influence of feedstock on the hydrotreating process is through a study of the hydrogen content (H/C atomic ratio) and molecular weight (carbon number) of the various feedstocks or products. It is also possible to use data for hydrogen usage in heavy feedstock processing where the relative amount of hydrogen consumed in the process can be shown to be dependent upon the sulfur content of the feedstock.

4.3.5 Gas Recycle

The recycle gas scheme is used in the HDS process to minimize physical losses of expensive hydrogen. HDS reactions require a high hydrogen partial pressure in the gas phase to maintain high desulfurization reaction rates and to suppress carbon laydown (catalyst deactivation). The high hydrogen partial pressure is maintained by supplying hydrogen to the reactors at several times the chemical hydrogen consumption rate. The majority of the unreacted hydrogen is cooled to remove hydrocarbons, recovered in the separator, and recycled for further utilization. Hydrogen is physically lost in the process by solubility in the desulfurized liquid hydrocarbon product, and from losses during the scrubbing or purging of hydrogen sulfide and light hydrocarbon gases from the recycle gas.

The reaction generally takes place in the vapor phase but, depending on the application, may be a mixed-phase reaction. The reaction products are cooled in a heat exchanger and led to a high-pressure separator where hydrogen gas is separated for recycling. Liquid products from the high-pressure separator flow to a low-pressure separator (stabilizer) where dissolved light gases are removed. The product may then be fed to a reforming or cracking unit, if desired.

4.3.6 Coke Formation

Attention must also be given to the coke mitigation aspects of hydrotreating as a preliminary treatment option of heavy feedstocks for other processes. Although the visbreaking process (Chapter 2) reduces the viscosity of heavy feedstocks and partially converts the residue to lighter hydrocarbons and coke, the process can also be used to remove the undesirable higher molecular weight polar constituents before sending the visbroken feedstock to a catalytic cracking unit. The solvent deasphalting process (Chapter 7) separates the higher value liquid product deasphalted oil using a light paraffinic solvent from low-value asphaltene-rich pitch stream. Various heavy feedstock hydrotreating (in fact, hydrocracking) processes (Chapter 5) in which the feedstock is processed under high temperature and pressure using a robust catalyst to remove sulfur, metals, condensed aromatic, or nitrogen and to increase the residue's hydrogen content to a desired degree are also available. However, an increased number of options are becoming available in which the heavy feedstock is first hydrotreated (under milder conditions to remove heteroatoms and to mitigate the effects of the asphaltene constituents and resin constituents) before sending the hydrotreated product to, for example, a fluid catalytic cracking unit.

In such cases, it is even more important that particular attention must be given to hydrogen management and promoting HDS and hydrodenitrogenation (HDN) (even fragmentation) of asphaltene and resin constituents thereby producing a product that may be suitable as a feedstock for catalytic cracking with reduced catalyst destruction. The presence of a material with good solvating power to assist in the hydrotreating process is preferred. In this respect it is worth noting the reappearance of donor solvent processing of heavy feedstocks (Vernon et al., 1984) that has its roots in the older hydrogen donor diluent visbreaking process (Carlson et al., 1958; Langer et al., 1962).

In addition, precautions need to be taken when unloading coked catalyst from the unit to prevent iron sulfide fires. The coked catalyst should be cooled to below 49°C (<120 °F) before removal, or dumped into nitrogen-blanketed bins where it can be cooled before further handling. Antifoam additives may be used to prevent catalyst poisoning from silicone carryover in the coker feedstock. There is a potential for exposure to hydrogen sulfide or hydrogen gas in the event of a release, or to ammonia should a sour-water leak or spill occur. Phenol also may be present if high-boiling-point feedstocks are processed.

4.4 CATALYST TECHNOLOGY

Typically, *hydrotreating catalysts* are usually cobalt plus molybdenum or nickel plus molybdenum (in the sulfide) forms, impregnated on an alumina base (Topsøe and Clausen, 1984). The hydrotreating operating conditions (1000–3000 psi hydrogen and about 455°C (850°F)) at which the desulfurization reactions proceed are invariably accompanied by small amounts of hydrogenation and hydrocracking, the extent of which depends on the nature of the feedstock and the severity of desulfurization.

HDS catalysts consist of metals impregnated on a porous alumina support. Almost all of the surface area is found in the pores of the alumina (200–300 m^2/g) and the metals are dispersed in a thin layer over the entire alumina surface within the pores. This type of catalyst does display an extensive catalytic surface for a small weight of catalyst. Cobalt (Co), molybdenum (Mo), and nickel (Ni) are the most commonly used metals for desulfurization catalysts. The catalysts are manufactured with the metals in an oxide state. In the active form they are in the sulfide state, which is obtained by sulfiding the catalyst either prior to use or with the feed during actual use. Any catalyst that exhibits hydrogenation activity will catalyze HDS to some extent. However, the Group VIB metals (chromium, molybdenum, and tungsten) are particularly active for desulfurization, especially when promoted with metals from the iron group (iron, cobalt, and nickel).

The increasing importance of HDS and HDN in petroleum processing in order to produce clean-burning fuels has led to a surge of research on the chemistry and engineering of heteroatom removal, with sulfur removal being the most prominent area of research. Most of the earlier works are focused (i) catalyst characterization by physical methods, (ii) low-pressure reaction studies of model compounds having relatively high reactivity, (iii) process development, or (iv) cobalt–molybdenum (Co–Mo) catalysts, nickel–molybdenum catalysts (Ni–Mo), or nickel–tungsten (Ni–W) catalysts supported on alumina, often doped by fluorine or phosphorus.

HDS and demetallization occur simultaneously on the active sites within the catalyst pore structure. Sulfur and nitrogen occurring in heavy feedstocks are converted to hydrogen sulfide and ammonia in the catalytic reactor and these gases are scrubbed out of the reactor effluent gas stream. The metals in the feedstock are deposited on the catalyst in the form of metal sulfides and cracking of the feedstock to distillate produces a laydown of carbonaceous material on the catalyst; both events poison the catalyst and activity or selectivity suffers. The deposition of carbonaceous material is a fast reaction that soon equilibrates to a particular carbon level and is controlled by hydrogen partial pressure within the reactors. On the other hand, metal deposition is a slow reaction that is directly proportional to the amount of feedstock passed over the catalyst.

The need to develop catalysts that can carry out deep HDS and deep HDN has become even more pressing in view of recent environmental regulations limiting the amount of sulfur and nitrogen emissions. The development of a new generation of catalysts to achieve this objective of low nitrogen and sulfur levels in the processing of different feedstocks presents an interesting challenge for catalyst development.

Basic nitrogen-containing compounds in a feed diminish the cracking activity of hydrocracking catalysts. However, zeolite catalysts can operate in the presence of substantial concentrations of ammonia, in marked contrast to silica–alumina catalysts, which are strongly poisoned by ammonia. Similarly, sulfur-containing compounds in a feedstock adversely affect the noble metal hydrogenation component of hydrocracking catalysts. These compounds are hydrocracked to hydrogen sulfide, which converts the noble metal to the sulfide form. The extent of this conversion is a function of the hydrogen and hydrogen sulfide partial pressures.

Removal of sulfur from the feedstock results in a gradual increase in catalyst activity, which returns almost to the original level. As with ammonia, the concentration of the hydrogen sulfide can be used to control precisely the chemical activity of the catalyst. Non-noble metal-loaded zeolite catalysts have an inherently different response to sulfur impurities since a minimum level of hydrogen sulfide is required to maintain the nickel–molybdenum and nickel–tungsten in the sulfide state.

HDN is more difficult to accomplish than HDS, but the relatively smaller amounts of nitrogen-containing compounds in conventional crude oil made this of lesser concern to refiners (Speight, 2007, 2011). However, the trend in refinery operations to heavier feedstocks, which are richer in nitrogen than the conventional feedstocks, has increased the awareness of refiners to the presence of nitrogen compounds in crude feedstocks. For the most part, however, HDS catalyst technology has been used to accomplish HDN (Topsøe and Clausen, 1984) although such catalysts are not ideally suited to nitrogen removal (Katzer and Sivasubramanian, 1979). However, in recent years, the limitations of HDS catalysts when applied to HDN have been recognized, and there are reports of attempts to manufacture catalysts more specific to nitrogen removal.

The poisoning effect of nitrogen can be offset to a certain degree by operation at a higher temperature. However, the higher temperature tends to increase the production of material in the methane (CH_4) to butane (C_4H_{10}) range and decrease the operating stability of the catalyst so that it requires more frequent regeneration. Catalysts containing platinum or palladium (approximately 0.5% wet) on a zeolite base appear to be somewhat less sensitive to nitrogen than are nickel catalysts, and successful operation has been achieved with feedstocks containing 40 ppm nitrogen. Such catalysts are also more tolerant of sulfur in the feed, which acts as a temporary poison, the catalyst recovering its activity when the sulfur content of the feed is reduced.

On catalysts such as nickel or tungsten sulfide on silica–alumina, isomerization does not appear to play any part in the reaction, as uncracked normal paraffin products from the feedstock tend to retain their normal structure. Extensive splitting produces large amounts of low-molecular-weight ($C_3 - C_6$) paraffins, and it appears that a primary reaction of paraffins is catalytic cracking followed by hydrogenation to form *iso*-paraffins. With catalysts of higher hydrogenation activity, such as platinum on silica–alumina, direct isomerization occurs. The product distribution is also different, and the ratio of low-molecular-weight to intermediate-molecular-weight paraffins in the breakdown product is reduced.

In addition to the chemical nature of the catalyst, the physical structure of the catalyst is also important in determining the hydrogenation and cracking capabilities, particularly for heavy feedstocks (Fischer and Angevine, 1986; Kang et al., 1988; van Zijll Langhout et al., 1980). When gas oils and heavy feedstocks are used, the feedstock is present as liquids under the conditions of the reaction. Additional feedstock and the hydrogen must diffuse through this liquid before reaction can take place at the interior surfaces of the catalyst particle.

At high temperatures, reaction rates can be much higher than diffusion rates and concentration gradients can develop within the catalyst particle. Therefore, the choice of catalyst porosity is an important parameter. When feedstocks are to be hydrocracked to liquefied petroleum gas and gasoline, pore diffusion effects are usually absent. High surface area (about 300 m²/g) and low- to moderate-porosity (from 12 D pore diameter with crystalline acidic components to 50 D or more with amorphous materials) catalysts are used. With reactions involving high-molecular-weight feedstocks, pore diffusion can exert a large influence, and catalysts with pore diameters greater than 80 D are necessary for more efficient conversion.

Catalyst operating temperature can influence reaction selectivity since the activation energy for hydrotreating reactions is much lower than for hydrocracking reaction. Therefore, raising the temperature in a heavy feedstock hydrotreater increases the extent of hydrocracking relative to hydrotreating, which also increases the hydrogen consumption.

Aromatic hydrogenation in petroleum refining may be carried out over supported metal or metal sulfide catalysts depending on the sulfur and nitrogen levels in the feedstock. For hydrorefining heavy feedstocks, which typically contain appreciable concentrations of sulfur and nitrogen, sulfided nickel—molybdenum (Ni—Mo), nickel—tungsten (Ni—W), or cobalt—molybdenum (Co—Mo) on alumina (γ-Al$_2$O$_3$) catalysts are generally used, whereas supported noble metal catalysts have been used for sulfur- and nitrogen-free feedstocks. Catalysts containing noble metals on Y-zeolites have been reported to be more sulfur tolerant than those on other supports (Jacobs, 1986).

Molybdenum sulfide (MoS_2), usually supported on alumina, is widely used in petroleum processes for hydrogenation reactions. It is a layered structure that can be made much more active by addition of cobalt or nickel (Topsøe et al., 1996). When promoted with cobalt sulfide (CoS), making what is called *cobalt–moly* catalysts, it is widely used in HDS processes. The nickel sulfide (NiS)-promoted version is used for HDN as well as HDS. The closely related tungsten compound (WS_2) is used in commercial hydrocracking catalysts. Other sulfides (iron sulfide, FeS, chromium sulfide, Cr_2S_3, and vanadium sulfide, V_2S_5) are also effective and used in some catalysts. A valuable alternative to the base metal sulfides is palladium sulfide (PdS). Although it is expensive, palladium sulfide forms the basis for several very active catalysts.

The life of a catalyst used to hydrotreat petroleum heavy feedstocks is dependent on the rate of carbon deposition and the rate at which organometallic compounds decompose and form metal sulfides on the surface. Several different metal complexes exist in the asphaltene fraction of the heavy feedstock and an explicit reaction mechanism of decomposition that would be a perfect fit for all of the compounds is not possible.

Different rates of reaction may occur with various types and concentrations of metallic compounds. For example, a medium-metal-content feedstock will generally have a lower rate of demetallization compared to high-metal-content feedstock. And, although individual organometallic compounds decompose according to both first- and second-order rate expressions, for reactor design, a second-order rate expression is applicable to the decomposition of heavy feedstock as a whole.

The current trend in hydroprocessing is the treatment of heavy sour feeds that contain compounds such as sulfur, nitrogen, aromatics, iron, and other undesirable components. These compounds pose significant problems with catalyst poisoning; however, developments are keeping pace with increased demand. In light of growing demand for ultra-low-sulfur diesel, light cycle oil hydrotreating is receiving much attention. Feeds such as these are typically high in heavy metals, which will require additional unit modifications and/or the installation of guard beds/reactors.

In addition, hydrotreating feedstocks prior to sending the feedstock to the fluid catalytic cracking units is another important focus and will continue to be important or even increase in importance. Many fluid catalytic cracking unit incorporate pretreaters (in the form of hydro-treating the feedstocks or guard beds/reactors) to meet their naphtha-gasoline sulfur requirements. Installation of such reactor units will nec-essarily increase as heavy feedstocks are incorporated into gas oils (fed to the fluid catalytic cracking unit) or become the sole feedstock for the catalytic cracking unit. Proven technology is available to remove sulfur, metals, and asphaltene content while converting an important part of the feed to lighter quality products. This technology will improve and operations will be varied to upgrade the following typical feedstocks: atmospheric and vacuum heavy feedstocks, tar sand bitu-men, deasphalter bottoms, and bio-feedstocks.

Thus, with the increasing focus to reduce sulfur content in fuels, the role of *desulfurization* in the refinery becomes more and more impor-tant. Currently, the process of choice is the hydrotreater, in which hydrogen is added to the fuel to remove the sulfur from the fuel. Some hydrogen may be lost to reduce the octane number of the fuel, which is undesirable. Because of the increased attention for fuel desulfuriza-tion various new process concepts are being developed with various claims of efficiency and effectiveness.

The major developments in desulfurization, embracing three main routes, are advanced hydrotreating (new catalysts, catalytic distillation, processing at mild conditions), reactive adsorption (type of adsorbent used, process design), and oxidative desulfurization (catalyst, process design) (Babich and Moulijn, 2003).

4.5 PROCESS OPTIONS

The major goal of *heavy feedstock hydrotreating* is to reduce yields of high-sulfur heavy fuel oil. This technology was originally developed to reduce the sulfur content of atmospheric residues to produce specifica-tion low sulfur heavy fuel oil. Changes in crude oil quality and product demand, however, have shifted the commercial importance of this technology to include pretreating conversion unit feedstocks to mini-mize catalyst replacement costs to reduce the yield and increase the quality of the by-product coke fraction. Although residue hydrotreaters are capable of processing feedstocks having a wide range of contami-nants, the feedstock's organometallic and asphaltene components typically determine its processability. Economics generally tend to limit residue hydrotreating applications to feedstocks with limitations (dictated by the process catalyst) on the content of nickel plus vanadium.

In many cases, application of hydrotreating technology to heavy feedstocks may also cause some cracking and, by inference, application of hydrocracking to heavy feedstocks will also cause desulfurization and denitrogenation. Rather than promote unnecessary duplication of the process description, two hydrotreating processes (resid desulfurization (RDS) and vacuum resid desulfurization (VRDS)) are described here together as subcategories of one process with the note that they are also amenable to hydrocracking operations.

4.5.1 Resid Desulfurization and Vacuum Resid Desulfurization

Heavy feedstock hydrotreating processes have two definite roles: (i) desulfurization to supply low-sulfur fuel oils and (ii) pretreatment of feed heavy feedstocks for heavy feedstock fluid catalytic cracking processes (Speight, 2007; Speight and Ozum, 2002). The main goal is to remove sulfur, metal, and asphaltene contents from heavy feedstocks and other heavy feedstocks to a desired level. On the other hand, the major goal of *heavy feedstock hydroconversion* is cracking of heavy feedstocks with desulfurization, metal removal, denitrogenation, and asphaltene conversion. Heavy feedstock hydroconversion processing offers production of kerosene and gas oil, and production of feedstocks for *hydrocracking, fluid catalytic cracking*, and petrochemical applications.

The RDS and VRDS process is (like the Residfining process, *q.v.*) a hydrotreating process that is designed to hydrotreat vacuum gas oil, atmospheric heavy feedstock, or vacuum heavy feedstock to remove sulfur metallic constituents while part of the feedstock is converted to lower boiling products. In the case of heavy feedstocks, the asphaltene content is reduced (Speight and Ozum, 2002).

The process consists of a once-through operation of hydrocarbon feed contacting graded catalyst systems designed to maintain activity and selectivity in the presence of deposited metals. Process conditions are designed for a 6-month to 1-year operating cycle between catalyst replacements. The process is ideally suited to produce feedstocks for heavy feedstock fluid catalytic crackers or delayed coking units to achieve minimal production of residual products.

The major product of the processes is a low-sulfur fuel oil and the amount of gasoline and middle distillates is maintained at a minimum to conserve hydrogen. The basic elements of each process are similar and consist of a once-through operation of the feedstock coming into contact with hydrogen and the catalyst in a downflow reactor that is designed to maintain activity and selectivity in the presence of deposited metals. Moderate temperatures and pressures are employed to reduce the incidence of hydrocracking and, hence, minimize production of low-boiling distillates (Speight, 2007; Speight and Ozum, 2002). The combination of a desulfurization step and a vacuum heavy feedstock desulfurizer (VRDS) is often seen as an attractive alternative to the atmospheric heavy feedstock desulfurizer (RDS). In addition, either the RDS or the VRDS option can be coupled with other processes (such as delayed coking, fluid catalytic cracking, and solvent deasphalting) to achieve optimal refining performance.

4.5.2 Residfining

The *Residfining* process is a catalytic fixed-bed process for the desulfurization and demetallization of heavy feedstocks (Speight, 2007; Speight and Ozum, 2002). The process can also be used to pretreat heavy feedstocks to suitably lower contaminant levels prior to catalytic cracking. In the process, liquid feed to the unit is filtered, pumped to pressure, preheated, and combined with treated gas prior to entering the reactors. A small guard reactor would typically be employed to prevent plugging/fouling of the main reactors. Provisions are employed to periodically remove the guard while keeping the main reactors online. The temperature rise associated with the exothermic reactions is controlled utilizing either a gas or liquid quench. A train of separators is employed to separate the gas and liquid products. The recycle gas is scrubbed to remove ammonia and H_2S. It is then combined with freshly made hydrogen before being reheated and recombined with fresh feed. The liquid product is sent to a fractionator where the product is fractionated.

The different catalysts allow other minor differences in operating conditions and peripheral equipment. Primary differences include the use of higher purity hydrogen makeup gas (usually 95% or greater), inclusion of filtration equipment in most cases, and facilities to upgrade the off-gases to maintain higher concentration of hydrogen in the recycle gas. Most of the processes utilize downflow operation over fixed-bed catalyst systems but exceptions to this are the H-Oil and

LC-Fining processes (which are predominantly conversion processes) that employ upflow designs and ebullating catalyst systems with continuous catalyst removal capability, and the Shell process (a conversion process), which may involve the use of a *bunker flow* reactor ahead of the main reactors to allow periodic changeover of catalyst.

The primary objective in most of the residue desulfurization processes is to remove sulfur with minimum consumption of hydrogen. Substantial percentages of nitrogen, oxygen, and metals are also removed from the feedstock. However, complete elimination of other reactions is not feasible and, in addition, hydrocracking, thermal cracking, and aromatic saturation reactions occur to some extent. Certain processes, that is, H-Oil (Chapter 5) using a single-stage or a two-stage reactor and LC-Fining (Chapter 5) using an expanded-bed reactor, can be designed to accomplish greater amounts of hydrocracking to yield larger quantities of lighter distillates at the expense of desulfurization.

Removal of nitrogen is much more difficult than removal of sulfur. For example, nitrogen removal may be only about 25–30% when sulfur removal is at a 75–80% level. Metals are removed from the feedstock in substantial quantities and are mainly deposited on the catalyst surface and exist as metal sulfides at processing conditions. As these deposits accumulate, the catalyst pores eventually become blocked and inaccessible, thus catalyst activity being lost.

Desulfurization of heavy feedstocks is considerably more difficult than desulfurization of distillates (including vacuum gas oil) because many more contaminants are present and very large, complex molecules are involved. The most difficult portion of feed in residue desulfurization is the asphaltene fraction that forms coke readily and it is essential that these large molecules be prevented from condensing with each other to form coke, which deactivates the catalyst. This is accomplished by selection of proper catalysts, use of adequate hydrogen partial pressure, and assuring intimate contact of the hydrogen-rich gases and oil molecules in the process design.

Finally, five of the most common approaches to upgrading hydrotreaters for heavy feedstocks (in order of increasing capital cost) are currently and will continue (at least for the next two decades) to be: (i) developing higher activity and more resilient catalysts, (ii) replacing reactor internals for increased efficiency, (iii) adding reactor capacity to accommodate heavy feedstocks and increase gasoline–diesel production, (iv) increasing hydrogen partial pressure, and (v) process design and hardware that are more specialized and focus on process schemes that effectively reduce hydrogen consumption.

4.5.3 Biodesulfurization

Another option under study for the processing of these crudes is biological upgrading, using fungus and bacteria selected and gathered in field operations and reproduced in the laboratory. There have been studies to characterize metabolic routes associated with the desulfurization and the elimination of other types of contaminants. Several institutions have identified microorganisms that work between 50°C and 65°C to degrade crude at atmospheric pressure, whilst the conventional processes require elevated temperatures and pressures. In microbial enhanced oil-recovery processes (Speight, 2007 and references cited therein), microbial technology is exploited in oil reservoirs to improve recovery. In the process, injected nutrients, together with indigenous or added microbes, promote in situ microbial growth and/or generation of products which mobilize additional oil and move it to producing wells through reservoir repressurization, interfacial tension/oil viscosity reduction, and selective plugging of the most permeable zones. Alternatively, the oil-mobilizing microbial products may be produced by fermentation and injected into the reservoir.

Biocatalyst desulfurization of petroleum distillates is one of a number of possible modes of applying biologically based processing to the needs of the petroleum industry (McFarland et al., 1998; Setti et al., 1999). In addition, *Mycobacterium goodii* has been found to desulfurize benzothiophene (Li et al., 2005). The desulfurization product was identified as α-hydroxystyrene. This strain appeared to have the ability to remove organic sulfur from a broad range of sulfur species in gasoline. When straight-run gasoline containing various organic sulfur compounds was treated with immobilized cells of *Mycobacterium goodii* for 24 h at 40°C (104°F), the total sulfur content significantly decreased, from 227 to 71 ppm at 40°C. Furthermore, when immobilized cells were incubated at 40°C (104°F) with *Mycobacterium goodii*, the sulfur content of the gasoline decreased from 275 to 54 ppm in two consecutive reactions.

A dibenzothiophene-degrading bacterial strain, *Nocardia* sp., was able to convert dibenzothiophene to 2-hydroxybiphenyl as the end metabolite through a sulfur-specific pathway (Chang et al., 1998). Other organic sulfur compounds, such as thiophene derivatives, thiazole derivatives, sulfides, and disulfides, were also desulfurized by *Nocardia* sp. When a sample in which dibenzothiophene was dissolved in hexadecane was treated with growing cells, the dibenzothiophene was desulfurized in approximately 80 h.

The soil-isolated strain microbe identified as *Rhodococcus erythropolis* can efficiently desulfurize benzonaphthothiophene (Yu et al., 2006). The desulfurization product was α-hydroxy-β-phenyl-naphthalene. Resting cells were able to desulfurize diesel oil (total organic sulfur, 259 ppm) after HDS and the sulfur content of diesel oil was reduced by 94.5% after 24 h at 30°C (86°F). Biodesulfurization of crude oils was also investigated and after 72 h at 30°C (86°F), 62.3% of the total sulfur content in Fushun crude oil (initial total sulfur content, 3210 ppm), and 47.2% of the sulfur in Sudanese crude oil (initial total sulfur, 1237 ppm) were removed. (See also Abbad-Andaloussi et al., 2003 and references cited therein.)

Heavy crude oil recovery facilitated by microorganisms was suggested in the 1920s and received growing interest in the 1980s in the form of microbial enhanced oil recovery. However, such projects have been slow to get under way although *in situ* bio-surfactant and biopolymer applications continue to garner interest (Van Hamme et al., 2003 and references cited therein). In fact, studies have been carried out on biological methods of removing heavy metals such as nickel and vanadium from petroleum distillate fractions, coal-derived liquid shale, bitumen, and synthetic fuels (Van Hamme et al., 2003 and references cited therein). However, further characterization on the biochemical mechanisms and bioprocessing issues involved in petroleum upgrading are required in order to develop reliable biological processes.

For upgrading options, the use of microbes has to show a competitive advantage of enzyme over the tried-and-true chemical methods prevalent in the industry. Currently, the range of reactions using microbes is large but is usually related to production of bioactive compounds or precursors. But the door is not closed and the issues of biodesulfurization and bio-upgrading remain open for the challenge of bulk petroleum processing. These drawbacks limit the applicability of this technology to specialty chemicals and steer it away from bulk petroleum processing.

Biotechnology has been widely used in bioremediation (Speight and Arjoon, 2012), but, in the case of upgrading, it is still necessary to go deeper in research and in extra laboratory test performance. It is expected that the use of biotechnological methods will eventually have a great impact on the development of crude upgrading. Biodesulfurization is, therefore, another technology to remove sulfur from the feedstock. However, several factors may limit the application of this technology. Many ancillary processes novel to petroleum refining would be needed, including a biocatalyst fermenter to regenerate the bacteria. The process is also sensitive to environmental conditions such as sterilization, temperature, and residence time of the biocatalyst. Finally, the process requires the existing hydrotreater to continue in operation to provide a lower sulfur feedstock to the unit and is more costly than conventional hydrotreating. Nevertheless, the limiting factors should not stop the investigations of the concept and work should be continued with success in mind.

Once the concept has been proven on the scale that a refiner would require, the successful microbial technology will most probably involve a genetically modified bacterial strain for (i) upgrading distillates and other petroleum fractions in refineries, (ii) upgrading crude petroleum upstream, and (iii) dealing with environmental problems that face industry, especially in areas related to spillage of crude oil and products. These developments are part of a wider trend to use bioprocessing to make products and do many of the tasks that are accomplished currently by conventional chemical processing. If commercialized for refineries, however, biologically based approaches will be at scales and with economic impacts beyond anything previously seen in industry.

In addition, the successful biodesulfurization process will, most likely, be based on naturally occurring aerobic bacteria that can remove organically bound sulfur in heterocyclic compounds without degrading the fuel value of the hydrocarbon matrix. Because of the susceptibility of bacteria to heat, the process will need to operate at temperatures and pressures close to ambient and also use air to promote sulfur removal from the feedstock.

REFERENCES

Abbad-Andaloussi, S., Warzywoda, M., Monot, F., 2003. Microbial desulfurization of diesel oils by selected bacterial strains. Revue Institut Français Du Pétrole 58 (4), 505–513.

Ancheyta, J., Speight, J.G., 2007. Hydroprocessing Heavy Oils and Heavy Feedstocks. CRC Press, Taylor & Francis Group, Boca Raton, FL.

Babich, I.V., Moulijn, J.A. 2003. Science and technology of novel processes for deep desulfuriza-tion of oil refinery streams: A Review. Fuel 82, 607–631.

Boening, L.G., McDaniel, N.K., Petersen, R.D., van Dreisen, R.P., 1987. Hydrocarbon Process. 66 (9), 59.

Carlson, C.S., Langer, A.W., Stewart, J., Hill, R.M., 1958. Chemical changes during thermal hydrocracking of Athabasca bitumen. Ind. Eng. Chem. 50, 1067.

Chang, J.H., Rhee, S.K., Chang, Y.K., Chang, H.N., 1998. Desulfurization of diesel oils by a newly isolated dibenzothiophene-degrading Nocardia sp. strain CYKS2. Biotechnol. Prog. 14 (6), 851–855.

Dolbear, G.E., 1998. In: Speight, J.G. (Ed.), Petroleum Chemistry and Refining. Taylor & Francis, Washington, DC (Chapter 7).

Fischer, R.H., Angevine, P.V., 1986. Appl. Catal. 27, 275.

Gary, J.H., Handwerk, G.E., Kaiser, M.J., 2007. Petroleum Refining: Technology and Economics, fifth ed. CRC Press, Taylor & Francis Group, Boca Raton, FL.

Howell, R.L., Hung, C., Gibson, K.R., Chen, H.C., 1985. Catalyst selection for residuum hydro-processing. Oil Gas J. 83 (30), 121–128.

Hsu, C.S., Robinson, P.R., 2006. Practical Advances in Petroleum Processing, vols. 1 and 2. Springer, New York, NY.

Jacobs, P.A., 1986. Metal clusters. In: Gates, B.C. (Ed.), Catalysis, Studies in Surface Science and Catalysis, 29. Elsevier, Amsterdam, p. 357.

Kang, B.C., Wu, S.T., Tsai, H.H., Wu, J.C., 1988. Appl. Catal. 45, 221.

Katzer, J.R., Sivasubramanian, R., 1979. Catal. Rev. Sci. Eng. 20, 155–208.

Khan, M.R., 1998. In: Speight, J.G. (Ed.), Petroleum Chemistry and Refining. Taylor & Francis, Washington, DC (Chapter 6).

Langer, A.W., Stewart, J., Thompson, C.E., White, H.Y., Hill, R.M., 1962. Ind. Eng. Chem. Proc. Design Dev. 1, 309.

Li, F., Xu, P., Feng, J., Meng, L., Zheng, Y., Luo, L., et al., 2005. Microbial desulfurization of gasoline in a Mycobacterium goodii X7B immobilized-cell system. Appl. Environ. Microbiol. 71 (1), 276–281.

McFarland, B.L., Boron, D.J., Deever, W., Meyer, J.A., Johnson, A.R., Atlas, R.M., 1998. Biocatalytic sulfur removal from fuels: applicability for producing low sulfur gasoline. Crit. Rev. Microbiol. 24, 99–147.

Meyers, R.A. (Ed.), 1997. Handbook of Petroleum Refining Processes. second ed. McGraw-Hill, New York, NY.

Reynolds, J.G., Beret, S., 1989. Effect of prehydrogenation on hydroconversion of Maya resid-uum. Fuel Sci. Technol. Int. 7, 165–186.

Setti, L., Farinelli, P., Di Martino, S., Frassinetti, S., Lanzarini, G., Pifferia, P.G., 1999. Appl. Microbiol. Biotechnol. 52, 111–117.

Speight, J.G., 2000. The Desulfurization of Heavy Oils and Heavy Feedstocks, second ed. Marcel Dekker Inc., New York, NY.

Speight, J.G., 2007. The Chemistry and Technology of Petroleum, fourth ed. CRC Press, Taylor & Francis Group, Boca Raton, FL.

Speight, J.G., 2011. The Refinery of the Future. Gulf Professional Publishing, Elsevier, Oxford, United Kingdom.

Speight, J.G., Arjoon, K.K., 2012. Bioremediation of Petroleum and Petroleum Products. Scrivener Publishing, Salem, MA.

Speight, J.G., Moschopedis, S.E., 1979. The production of low-sulfur liquids and coke from Athabasca bitumen. Fuel Process. Technol. 2, 295.

Speight, J.G., Ozum, B., 2002. Petroleum Refining Processes. Marcel Dekker Inc., New York, NY.

Topsøe, H., Clausen, B.S., 1984. Catal. Rev. Sci. Eng. 26, 395.

Topsøe, H., Clausen, B.S., Massoth, F.E., 1996. In: Anderson, J.R., Boudart, M. (Eds.), Hydrotreating Catalysis in Catalysis — Science and Technology, vol. 11. Springer Verlag, Berlin, Germany.

Topsøe, H., Egeberg, R.G., Knudsen, K.G., 2004. Future challenges of hydrotreating catalyst technology. Div. Fuel Chem., Am. Chem. Soc. 49 (2), 568—569 (Preprints).

Van Hamme, J.D., Singh, A., Ward, O.P., 2003. Recent advances in petroleum microbiology. Microbiol. Mol. Biol. Rev. 67 (4), 503—549.

Van Zijll Langhout, W.C., Ouwerkerk, C., Pronk, K.M.A., 1980. New process hydrotreats metal-rich feedstocks. Oil Gas J. 78 (48), 120—126.

Vernon, L.W., Jacobs, F.E., Bauman, R.F., 1984. Process for converting petroleum residuals. United States Patent 4,425,224. January 10.

Yu, B., Xu, P., Shi, Q., Ma, C., 2006. Deep desulfurization of diesel oil and crude oils by a newly isolated *Rhodococcus erythropolis* strain. Appl. Environ. Microbiol. 72, 54—58.

CHAPTER 5

Hydrocracking

5.1 INTRODUCTION

The major differences between *hydrocracking* and *hydrotreating* (Chapter 4) are the time at which the feedstock remains at reaction temperature and the extent of the decomposition of the nonheteroatom constituents and products. The lower limits of hydrocracking conditions may overlap with the upper limits of hydrotreating conditions. Where the reaction conditions overlap, feedstocks to be hydrocracked will generally be exposed to the reactor temperature for longer periods; hence the reason why hydrocracking conditions may be referred to as (relatively) severe (Gary et al., 2007; Hsu and Robinson, 2006; Speight, 2000, 2007; Speight and Ozum, 2002).

Heavy feedstocks—which require more energy-intensive processing than conventional crude oil—will contribute a growing fraction of fuels production. As existing reserves of conventional oil are depleted and there is greater worldwide competition for premium (e.g., light, sweet) crude oil, refineries will increasingly utilize heavy oil, sour crude oil, and tar sand bitumen to meet demand. As a result, it is apparent that the conversion of heavy feedstocks requires new lines of thought to develop suitable processing scenarios (Speight, 2011a). This has brought about, and will continue to bring about in the refinery of the future, the evolution of processing schemes that accommodate the heavier feedstocks (Boening et al., 1987; Hsu and Robinson, 2006; Khan and Patmore, 1998; Speight, 2007, 2011a; Speight and Ozum, 2002).

5.2 PROCESS TYPES

As with hydrotreating units, many different hydrocracking unit designs are marketed, and they all work along the same principle—all processes use the reaction of hydrogen with the hydrocarbon feedstock to produce hydrogen sulfide and a desulfurized hydrocarbon product (Gary et al., 2007; Hsu and Robinson, 2006; Speight, 2007). This also invokes the concept of different process types (naming is usually based on the types of reactor employed). Each reactor has its own merits and, provided the choice is made according to the feedstock properties has an excellent chance of producing the desired products.

The reaction temperature is typically at the upper end of the range 290–445°C (550–850°F) with a hydrogen gas pressure at the upper end of the range 150 and 3000 psi—the higher temperature maximizes cracking reactions. In some designs, the feedstock is heated and then mixed with the hydrogen rather than the option of passing moderately heated feedstock (i.e., feedstock not at the full reactor temperature) and moderately heated hydrogen (i.e., hydrogen not at the full reactor temperature) into the reactor. The gas mixture is led over a catalyst bed of metal oxides (most often cobalt or molybdenum oxides on different metal carriers). The catalysts help the hydrogen to react with sulfur and nitrogen to form hydrogen sulfides (H_2S) and ammonia. The reactor effluent is then cooled, and the oil feed and gas mixture is then separated in a stripper column. Part of the stripped gas may be recycled to the reactor.

Generally, with few exceptions, hydrocracking reactors fall into the same groups as those used for hydrotreating: (1) downflow fixed-bed reactor, upflow expanded-bed reactor, and a demetallization reactor. In addition, process variables focus on (1) process flow, feedstock properties, reaction temperature, hydrogen partial pressure, gas recycle, and coke formation. Since these items are discussed in detail elsewhere (Chapter 4), it is unnecessary to repeat the descriptions. Any differences will be presented in the section dealing with process options (presented below). However, there are several issues that need to be looked at here in relation to focusing on the future of the hydrocracking process.

For the heavy feedstocks, which will increase in amount in terms of hydrocracking feedstocks, reactor designs focus on online catalyst addition and withdrawal. Fixed-bed designs have suffered from (1) mechanical inadequacy when used for the heavier feedstocks and (2) short catalyst lives—6 months or less—even though large catalyst volumes are used (LHSV typically of 0.5–1.5). Refiners will attempt to overcome these shortcomings by innovative designs, allowing better feedstock flow and catalyst utilization or online catalyst removal. For example, processes that focus on on-stream catalyst replacement (OCR) in which a lead, moving bed reactor is used to demetallize heavy feedstock(s) ahead of the fixed-bed hydrocracking reactor will continue to be of interest and see further use.

The use of ebullating bed technologies was first introduced in the 1960s in an attempt to overcome problems of catalyst aging and poor distribution in fixed-bed designs. Hydrogen and feed enter at the bottom of the reactor, thereby expanding the catalyst bed. Although catalyst performance can be kept constant because catalyst can be replaced online, ebullition results in a back mixed reactor; therefore, desulfurization and hydroconversion are less than obtainable in a fixed-bed unit. Currently, in order to limit coking, most commercial ebullating bed units operate in the 70–85% desulfurization range and for 50–70% v/v, heavy feedstock conversion.

Development work will continue and ebullating bed units will see more use and have a greater impact of heavy feedstock conversion operations. Improvements include for example (1) second generation catalyst technology, which will allow higher conversion to a stable product, (2) catalyst rejuvenation, which allows spent catalyst to be reused to a greater extent than current operations allow, and (3) new reactor designs raising single train size throughput.

Slurry-phase hydrocracking of heavy oil and the latest development of dispersed catalysts present strong indications that such technologies will play a role in future refineries. Catalysts for slurry-phase hydrocracking of heavy oil have undergone two development phases: (1) heterogeneous solid powder catalysts, which have low catalytic activity and will produce a large number of solid particles in bottom oil making the catalyst difficult to dispose and utilize, and (2) homogeneous dispersed catalysts, which are divided into water-soluble dispersed catalysts and oil-soluble dispersed catalysts (Zhang et al., 2007). Dispersed catalysts are highly dispersed and have greater surface area to volume ratio. Therefore, they show high catalytic activity and good performance. They are desirable catalysts for slurry-phase hydrocracking of heavy oil and will be used more prominently in future hydrocracking operations.

In spite of the numerous process design variations, process design innovations and hardware innovations will continue. Although conventional (high-pressure) hydrocracking will still be used to address the need to produce more gasoline and diesel using moderate-pressure hydrocracking (where control of the reaction chemistry is more possible), another approach is to introduce a mild hydrocracking unit upstream of the fluid catalytic cracking unit to maintain that unit at full capacity. Alternatively, more refiners will turn to two-stage recycle hydrocracking (TSR hydrocracking) and reverse-staging configurations.

Another central focus will be the reduction of reducing hydrogen consumption while maintaining product quality. Catalysts that can withstand organic nitrogen contamination are being developed for lower-cost, single-stage units. The addition of metal traps upstream of the hydroprocessing unit is a solution to protect highly active catalyst from high metals feeds that will see wider application.

Furthermore, biomass gasification and Fischer—Tropsch synthesis conversion are very likely to be a part of the future refineries as part of next-generation biofuels developments (Speight, 2011b); refiners will need to monitor closely the latest refinery-related advances as well as future directions in biomass processing. In particular, the response of the refining industry to opportunities for processing heavy feedstocks, mandated biofuels usage, and requirements to comply with carbon dioxide will need to be addressed. Furthermore, gasification with carbon capture and the use of biomass as feedstock should help refiners meet emissions reduction requirements for carbon dioxide.

5.3 CATALYST TECHNOLOGY

The hydrocracking process employs high-activity catalysts that produce a significant yield of light products. In addition to the increased hydrocracking activity of the catalyst, percentage desulfurization and denitrogenation at start-of-run conditions are also substantially increased.

Hydrocracking catalysts typically contain separate hydrogenation and cracking functions. Palladium sulfide and promoted group VI sulfides (nickel molybdenum or nickel tungsten) provide the hydrogenation function. These active compositions saturate aromatics in the feed, saturate olefins formed in the cracking, and protect the catalysts from poisoning by coke. Zeolites or amorphous silica—alumina provide the cracking functions. The zeolites are usually of type Y (faujasite), ion exchanged to replace sodium with hydrogen, and make up

25–50% of the catalysts. Pentasils (silicalite or ZSM-5) may be included in dewaxing catalysts.

The deposition of coke and metals on to the catalyst diminish the cracking activity of hydrocracking catalysts. Basic nitrogen plays a major role because of the susceptibility of such compounds to the catalyst and their predisposition to form coke (Speight, 2000; Speight and Ozum, 2002). The reactions of hydrocracking require a dual-function catalyst with high cracking and hydrogenation activities (Ho, 1988; Katzer and Sivasubramanian, 1979). The catalyst base, such as acid-treated clay, usually supplies the cracking function; alumina or silica–alumina is used to support the hydrogenation function supplied by metals, such as nickel, tungsten, platinum, and palladium. These highly acid catalysts are very sensitive to nitrogen compounds in the feed, which break down the conditions of reaction to give ammonia and neutralize the acid sites. As many heavy gas oils contain substantial amounts of nitrogen (up to approximately 2500 ppm), a purification stage is frequently required. Denitrogenation and desulfurization can be carried out using cobalt–molybdenum or nickel–cobalt–molybdenum on alumina or silica–alumina.

Hydrocracking catalysts, such as nickel (5% by weight) or silica–alumina, work best on feedstocks that have been hydrofined to low nitrogen and sulfur levels. The nickel catalyst then operates well at 350–370°C (660–700°F) and a pressure of about 1500 psi to give good conversion of feed to lower-boiling liquid fractions with minimum saturation of single-ring aromatics and a high iso-paraffin to n-paraffin ratio in the lower molecular weight paraffins.

The poisoning effect of nitrogen can be offset to a certain degree by operation at a higher temperature. However, the higher temperature tends to increase the production of material in the methane (CH_4) to butane (C_4H_{10}) range and decrease the operating stability of the catalyst so that it requires more frequent regeneration. Catalysts containing platinum or palladium (approximately 0.5% wet) on a zeolite base appear to be somewhat less sensitive to nitrogen than are nickel catalysts, and successful operation has been achieved with feedstocks containing 40 ppm nitrogen. This catalyst is also more tolerant of the feed containing sulfur, which acts as a temporary poison, and the catalyst recovers its activity when the sulfur content of the feed is reduced.

On such catalysts such as nickel or tungsten sulfide on silica–alumina, isomerization does not appear to play any part in the reaction, as uncracked normal paraffin compounds from the feedstock tend to retain their normal structure. Extensive splitting produces large amounts of low-molecular-weight ($C_3 - C_6$) paraffins, and it appears that a primary reaction of paraffins is catalytic cracking followed by hydrogenation to form *iso*-paraffins. With catalysts of higher hydrogenation activity, such as platinum on silica–alumina, direct isomerization occurs. The product distribution is also different, and the ratio of low- to intermediate-molecular-weight paraffins in the breakdown product is reduced.

In addition to the chemical nature of the catalyst, the physical structure of the catalyst is important in determining the hydrogenation and cracking capabilities, particularly for heavy feedstocks (Kang et al., 1988; Kobayashi et al., 1987; van Zijll Langhout et al., 1980). When gas oils and heavy feedstocks are used, the feedstock is present as liquids under the conditions of the reaction. Additional feedstock and the hydrogen must diffuse through this liquid before reaction can take place at the interior surfaces of the catalyst particle.

At high temperatures, reaction rates can be much higher than diffusion rates and concentration gradients can develop within the catalyst particle. Therefore, the choice of catalyst porosity is an important parameter. When feedstocks are to be hydrocracked to liquefied petroleum gas and gasoline, pore diffusion effects are usually absent. High surface area (about $300 \, m^2/g$) and low- to moderate-porosity (from 12D pore diameter with crystalline acidic components to 50D or more with amorphous materials) catalysts are used. With reactions involving high-molecular-weight feedstocks, pore diffusion can exert a large influence, and catalysts with pore diameters greater than 80D are necessary for more efficient conversion.

Catalyst operating temperature can influence reaction selectivity since the activation energy for hydrotreating reactions is much lower than for hydrocracking reaction. Therefore, raising the temperature in a heavy feedstock hydrotreater increases the extent of hydrocracking relative to hydrotreating, which also increases the hydrogen consumption.

Molybdenum sulfide (MoS_2), usually supported on alumina, is widely used in petroleum processes for hydrogenation reactions. It is a layered structure that can be made much more active by addition of cobalt or nickel. When promoted with cobalt sulfide (CoS), making what is called *cobalt-moly* catalysts, it is widely used in hydrodesulfurization (HDS) processes. The nickel sulfide (NiS)-promoted version is used for hydrodenitrogenation (HDN) as well as HDS. The closely related tungsten compound (WS_2) is used in commercial hydrocracking catalysts. Other sulfides (iron sulfide, FeS, chromium sulfide, Cr_2S_3, and vanadium sulfide, V_2S_5) are also effective and used in some catalysts. A valuable alternative to the base metal sulfides is palladium sulfide (PdS). Although it is expensive, palladium sulfide forms the basis for several very active catalysts.

Clay minerals have been used as cracking catalysts particularly for heavy feedstocks (Ancheyta and Speight, 2007; Speight, 2007) and their use has also been explored in relation to the demetallization and upgrading of heavy crude oil (Rosa-Brussin, 1995). The results indicated that the catalyst prepared was mainly active toward demetallization and conversion of the heaviest fractions of crude oils.

The cracking reaction results from attack of a strong acid on a paraffinic chain to form a carbonium ion (a carbon cation, e.g., R^+) (Dolbear, 1998). Strong acids come in two fundamental types, Brønsted and Lewis acids. *Brønsted acids* are the familiar proton-containing acids; *Lewis acids* are a broader class including inorganic and organic species formed by positively charged centers. Both kinds have been identified on the surfaces of catalysts; sometimes both kinds of sites occur on the same catalyst. The mixture of Brønsted and Lewis acids sometimes depends on the level of water in the system.

Examples of Brønsted acids are the familiar proton-containing species such as sulfuric acid (H_2SO_4). Acidity is provided by the very active hydrogen ion (H^+), which has a very high positive charge density. It seeks out centers of negative charge such as the pi electrons in aromatic centers. Such reactions are familiar to organic chemistry students, who are taught that bromination of aromatics takes place by attack of the bromonium ion (Br^+) on such a ring system. The proton in strong acid systems behaves in much the same way, adding to the pi electrons and then migrating to a site of high electron density on one of the carbon atoms.

In reactions with feedstock constituents, both Lewis and Brønsted acids can catalyze cracking reactions. For example, the proton in Brønsted acids can add to an olefinic double bond to form a carbon cation. Similarly, a Lewis acid can abstract a hydride from the corresponding paraffin to generate the same intermediate (Dolbear, 1998). Although these reactions are written to show identical intermediates in the two reactions, in real catalytic systems, the intermediates would be different.

Zeolite catalysts have also found use in the refining industry during the last two decades. Like the silica–alumina catalysts, zeolites also consist of a framework of tetrahedral form usually with a silicon atom or an aluminum atom at the center (Occelli and Robson, 1989).

Zeolite catalysts have also shown remarkable adaptability to the refining industry. For example, the resistance to deactivation of the type Y zeolite catalysts containing either noble or non-noble metals is remarkable, and catalyst life of up to 7 years has been obtained commercially in processing heavy gas oils in the Unicracking-JHC processes. Operating life depends on the nature of the feedstock, the severity of the operation, and the nature and extent of operational upsets. Gradual catalyst deactivation in commercial use is counteracted by incrementally raising the operating temperature to maintain the required conversion per pass. The more active a catalyst, the lower is the temperature required. When processing for gasoline, lower operating temperatures have the additional advantage that less of the feedstock is converted to *iso*-butane.

While zeolites provided a breakthrough that allowed catalytic hydrocracking to become commercially important, continued advances in the manufacture of amorphous silica–alumina made these materials competitive in certain kinds of applications. This was important, because patents controlled by Unocal and Exxon dominated the application of zeolites in this area.

Typical catalysts of this type contain 60–80% w/w of silica alumina, with the remainder being the hydrogenation component. The compositions of these catalysts are closely held secrets. Over the years, broad ranges of silica/alumina molar ratios have been used in various cracking applications, but silica is almost always in excess for high acidity and stability. A typical level might be 25% w/w alumina (Al_2O_3).

Hydrocarbons, especially aromatic hydrocarbons as found in heavy feedstocks, can react in the presence of strong acids to form coke. This coke is a complex polynuclear aromatic material that is low in hydrogen. Coke can deposit on the surface of a catalyst, blocking access to the active sites and reducing the activity of the catalyst. Coke poisoning is a major problem in fluid catalytic cracking catalysts, where coked catalysts are circulated to a fluidized bed combustor to be regenerated. In hydrocracking, coke deposition is virtually eliminated by the catalyst's hydrogenation function.

However, the product referred to as *coke* is not a single material. The first products deposited are tarry deposits that can, with time and temperature, continue to polymerize. Acid catalyzes these polymerizations. The stable product would be graphite, with very large aromatic sheets and no hydrogen. This product forms only with very high-temperature aging, far more severe than that found in a hydrocracker. The graphitic material is both more thermodynamically stable and less kinetically reactive. This kinetic stability results from the lack of easily hydrogenated functional groups.

Catalysts carrying coke deposits can be regenerated by burning off the accumulated coke. This is done by service in rotary or similar kilns rather than leaving catalysts in the hydrocracking reactor, where the reactions could damage the metals in the walls. Removing the catalysts also allows inspection and repair of the complex and expensive reactor internals, discussed below. Regeneration of a large catalyst charge can take weeks or months, so refiners may own two catalyst loads, one in the reactor, and one regenerated and ready for reload.

Precious metal catalysts, particularly catalysts incorporating platinum or platinum and palladium, are used in the latter stages of deep-desulfurization process. They have excellent performance in hydrogenation of monocyclic aromatic hydrocarbons and are likely to become more and more important in the years ahead and refiners seek to hydrocrack higher amounts of the heavy feedstocks. Current efforts are seeking to produce such catalysts with increased hydrogenation activity as well as resistance to sulfur and nitrogen poisoning while balancing these characteristics. In addition, work is being done to assign appropriate cracking activity to match specific applications using inorganic (composite) oxides, such as amorphous alumina, silica alumina, or crystalline silica alumina, as carriers and for optimization of the amount of the precious metals and highly dispersed metal impregnation methods.

Catalysts used in heavy feedstock upgrading processes typically use an association of several kinds of catalysts, each of them playing a specific and complementary role (Kressman et al., 1998). The first major function to be performed is hydrodemetallization (HDM). Therefore, the HDM catalyst must desegregate asphaltene constituents and remove as much metal (nickel and vanadium) as possible. One catalyst in particular has been developed by optimizing the support pore structure and acidity (Toulhoat et al., 1990). This catalyst allows a uniform distribution of metals deposited and therefore a high metal retention capacity is reached. A specific HDS catalyst can be placed downstream of the HDM catalyst and the main function of such positioning is to desulfurize the already deeply demetallized feedstock as well as to reduce coke precursors. Thus, the main function of the HDS catalyst is not the same as that of the HDM catalyst. In addition, for fixed-bed processes, swing guard reactors may be used to improve the protection of downstream catalysts and increases the unit cycle length. For example, the Hyvahl process (*q.v.*) includes two swing guard reactors followed by conventional HDM and HDS reactors (DeCroocq, 1997). The HDM catalyst in the guard reactors may be replaced during unit operation and the total catalyst amount is replaced at the end of a cycle.

Catalyst improvements will continue to (i) improve hydrocracking activity, (ii) improve HDS activity, (iii) reduce catalytic deactivation, and (iv) increase cycle length. The developments in nonprecious metal catalysts, hetero-polyanions to improve metal dispersions, beta zeolite, and the acid-cracking-based formulations of highly active hydrocracking catalysts, have already added (and will continue to add) flexibility in the operations of hydroprocessing units. New formulations that employ amorphous silica–alumina supports and de-aluminated Y-zeolites will be readily available and offer high activity with high stability. These designs allow for lower operating pressures, increased run length, and higher gasoline and diesel yields.

Finally, improved catalysts based on a better understanding of asphaltene chemistry will be a focus of catalyst manufacture—to hydrocrack asphaltene constituents without serious deleterious effect on the catalyst. In this respect, the reemergence of the former CANMET process as the Uniflex process is a major step forward since processing Athabasca bitumen using the CANMET process gave good conversion to distillates and the iron-based catalyst also acted as a scavenger for coke formers—other options include the addition of metal oxides as scavengers for coke formers and sulfur.

Investment in new technologies is on the move and, in fact, the application of hydrogen addition technologies to heavy feedstocks is increasing.

5.4 PROCESSES OPTIONS

The hydrocracking process in which hydrogen is used is an attempt to *stabilize* the reactive fragments produced during the cracking, thereby decreasing their potential for recombination to heavier products and ultimately to coke. The choice of processing schemes for a given hydrocracking application depends upon the nature of the feedstock as well as the product requirements. The process can be simply illustrated as a single-stage or as a two-stage operation (Figure 5.1).

Generally, the refinery utilizes one of three options for the process. Thus, depending on the feedstock being processed and the type of plant design employed (*single-stage* or *two-stage*), flexibility can be provided to vary product distribution among the following principal end products.

The single-stage process can be used to produce gasoline but is more often used to produce middle distillate from heavy vacuum gas oils. The two-stage process was developed primarily to produce high yields of gasoline from straight-run gas oil, and the first stage may actually be a purification step to remove sulfur-containing (as well as nitrogen-containing) organic materials. In terms of sulfur removal, it appears that nonasphaltene sulfur in nonasphaltene constituents may be removed before the more refractory sulfur in asphaltene constituents (Speight, 2007) thereby requiring thorough desulfurization. This is a good reason for processes to use an extinction-recycling technique to maximize desulfurization and the yields of the desired product. Significant conversion of heavy feedstocks can be accomplished by hydrocracking at high severity (Howell et al., 1985). For some applications, the products boiling up to 340°C (650°F) can be blended to give the desired final product.

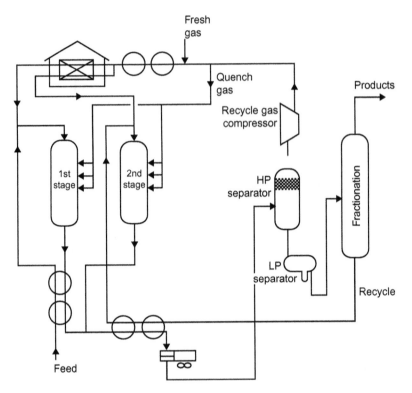

Figure 5.1 A single-stage or two-stage (optional) hydrocracking/hydrotreating unit. OSHA Technical Manual, Section IV, Chapter 2: Petroleum Refining Processes. http://www.osha.gov/dts/osta/otm/otm_iv/otm_iv_2.html.

Many hydrocracking units use fixed beds of catalyst with downflow of reactants. However for heavier feedstocks, the H-Oil process and the LC-Fining processes employ an ebullient bed reactor in which the beds of particulate catalyst are maintained in an ebullient or fluidized condition with upflowing reactants.

The major goal of *heavy feedstock hydroconversion* is cracking of heavy feedstocks with desulfurization, metal removal, denitrogenation, and conversion of asphaltene constituents. Heavy feedstock hydroconversion process offers production of kerosene and gas oil, and production of feedstocks for *hydrocracking, fluid catalytic cracking*, and petrochemical applications. The processes that follow are listed in alphabetical order with no preference in mind.

5.4.1 Asphaltenic Bottom Cracking Process
The asphaltenic bottom cracking (ABC) process can be used for distillate production (Table 5.1), HDM, asphaltene cracking, and moderate

Table 5.1 Feedstock and Product Adapt for the ABC Process			
Feedstock	Arabian Light Vacuum Resid	Arabian Heavy Vacuum Resid	Cerro Negro Vacuum Resid
API	7.0	5.1	1.7
Sulfur % w/w	4.0	5.3	4.3
Carbon heavy feedstock % w/w	20.8	23.3	23.6
C7-asphaltenes	7.0	13.1	19,819.8
Nickel, ppm	223.0	52.0	150.0
Vanadium, ppm	76.0	150.0	640.0
Products			
Naphtha, vol.%	6.5	7.7	15.1
API	57.2	57.2	54.7
Distillate, vol.%	16.0	19.8	21.3
API	34.2	34.2	32.5
Vacuum gas oil, vol.%	34.3	38.1	32.8
API	24.7	21.6	15.4
Sulfur % w/w	0.2	1.7	0.5
Vacuum heavy feedstock, vol.%	46.2	37.9	34.7
API	10.6	7.8	<0.0
Sulfur % w/w%	0.6	1.7	2.2
Carbon heavy feedstock % w/w	13.6	26.5	13.6
C7-asphaltenes % w/w			
Nickel, ppm	9.0	45.0	117.0
Vanadium, ppm	11.0	75.0	371.0
Conversion	55.0	60.0	60.0
Source: Speight (2007).			

HDS, as well as providing sufficient resistance to coke fouling and metal deposition using heavy feedstocks fixed catalyst beds.

The process can be combined with: (i) solvent deasphalting for complete or partial conversion of the heavy feedstock or (ii) HDS to promote the conversion of heavy feedstock, to treat feedstock with high metals, and to increase catalyst life, or (iii) hydrovisbreaking to attain high conversion of heavy feedstock with favorable product stability.

Within the process, the feedstock is pumped up to the reaction pressure and mixed with hydrogen. The mixture is heated to the reaction temperature in the charge heater after heat exchange and fed into the reactor.

In the reactor, HDM and subsequent asphaltene cracking with moderate HDS take place simultaneously under conditions similar to heavy feedstock HDS. The reactor effluent gas is cooled, cleaned up, and recycled to the reactor section, while the separated liquid is distilled into distillate fractions and vacuum heavy feedstock which is further separated by deasphalting (Chapter 6) into deasphalted oil and asphalt using butane or pentane.

In case of ABC–HDS catalyst combination, the ABC catalyst is placed upstream of the HDS catalyst and can be operated at a higher temperature than with the HDS catalyst under conventional heavy feedstock HDS conditions. In the VisABC process, a soaking drum is provided after the heater, when necessary. Hydrovisbroken oil is first stabilized by the ABC catalyst through hydrogenation of coke precursors and then desulfurized by the HDS catalyst.

5.4.2 H-Oil Process

The H-Oil process (Speight, 2007; Speight and Ozum, 2002) is a catalytic process that uses a single-stage, two-stage, or three-stage ebullated-bed reactor in which, during the reaction, considerable hydrocracking takes place (Figure 5.2). The process is designed for hydrogenation of heavy feedstocks and other high feedstocks in an ebullated-bed reactor to produce upgraded petroleum products (Speight, 2007; Speight and Ozum, 2002). The process is able to convert all types of feedstocks to distillate products as well as to desulfurize and demetallize heavy feedstocks for feed to coking units or heavy feedstock fluid catalytic cracking units, for production of low sulfur fuel oil, or for production to asphalt blending. A modification of the H-Oil process (the *Hy-C Cracking process*) converts high-boiling distillates to middle distillates and kerosene (Table 5.2) (Speight and Ozum, 2002).

A wide variety of process options can be used with the H-Oil process depending on the specific operation. In all cases, a catalytic ebullated-bed reactor system is used to provide an efficient hydroconversion. The system insures uniform distribution of liquid, hydrogen-rich gas,

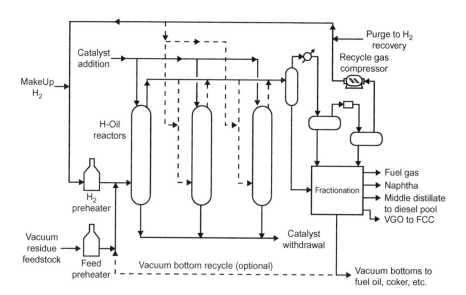

Figure 5.2 H-Oil process. Speight (2007).

and catalyst across the reactor. The ebullated-bed system operates under essentially isothermal conditions, exhibiting little temperature gradient across the bed (Kressmann et al., 2000). The heat of reaction is used to bring the feed oil and hydrogen up to reactor temperature.

Within the process, the feedstock (a vacuum heavy feedstock) is mixed with recycle vacuum heavy feedstock from downstream fractionation, hydrogen-rich recycle gas, and fresh hydrogen. This combined stream is fed into the bottom of the reactor whereby the upward flow expands the catalyst bed. The mixed vapor liquid effluent from the reactor goes to either flash drum for phase separation or the next reactor. A portion of the hydrogen-rich gas is recycled to the reactor. The product oil is cooled and stabilized and the vacuum heavy feedstock portion is recycled to increase conversion.

A catalyst of small particle size can be used, giving efficient contact among gas, liquid, and solid with good mass and heat transfer. Part of the reactor effluent is recycled back through the reactors for temperature control and to maintain the requisite liquid velocity. The entire bed is held within a narrow temperature range, which provides essentially an isothermal operation with an exothermic process. Because of the movement of catalyst particles in the liquid–gas medium,

Feedstock	Arabian Medium Vacuum Resid	Arabian Medium Vacuum Resid	Athabasca Bitumen
Table 5.2 Feedstock and Product Data for the H-Oil Process			
API gravity	4.9	4.9	8.3
Sulfur % w/w	5.4	5.4	4.9
Nitrogen % w/w			0.5
Carbon residue % w/w			
Metals, ppm	128.0	128.0	
Ni			
V			
Resid (> 525°C, >975°F) % w/w			50.3
Products % w/w**			
Naphtha (C5—204°C, 400°F)	17.6	23.8	16.0
Sulfur % w/w			1.0
Distillate (204−343°C, 400−650°F)	22.1	36.5	43.0
Sulfur % w/w			2.0
Vacuum gas oil (343−534°C, 650−975°F)	34.0	37.1	26.4
Sulfur % w/w			3.5
Heavy feedstock (> 534°C, >975°F)	33.2	9.5	16.0
Sulfur % w/w			5.7
*% conversion			
**% desulfurization			
Source: Speight (2007).			

deposition of tar and coke is minimized and fine solids entrained in the feed do not lead to reactor plugging. The catalyst can also be added into and withdrawn from the reactor without destroying the continuity of the process. The reactor effluent is cooled by exchange and separates into vapor and liquid. After scrubbing in a lean oil absorber, hydrogen is recycled and the liquid product is either stored directly or fractionated before storage and blending.

H-Oil and LC-Fining technologies are often practiced commercially at about the 85% conversion level.

5.4.3 Hydrovisbreaking Process

By way of recall, visbreaking (Chapter 2) is a relatively simple process that consists of a large furnace heating the feedstock to the range of

450–500°C (840–930°F) at an operating pressure of about 140 psi (965 kPa). The residence time in the furnace is carefully limited to prevent much of the reaction from taking place and clogging the furnace tubes. The heated feed is then charged to a coil or a reaction chamber that is maintained at a pressure high enough to permit cracking of the large molecules but restrict coke formation. From the reaction area, the process fluid is cooled (quenched) to inhibit further cracking and then charged to a distillation column for separation into components.

Visbreaking units typically convert about 15% of the feedstock to naphtha and diesel oils and produce a lower-viscosity residual fuel. Thermal cracking units provide more severe processing and often convert as much as 50–60% of the incoming feed to naphtha and light diesel oil.

Briefly, the *hydrovisbreaking* process (HYCAR process) is a noncatalytic process conducted under similar conditions to visbreaking (Chapter 2) and involves treatment with hydrogen under mild conditions. The presence of hydrogen leads to more stable products (lower *flocculation threshold*) than can be obtained with straight visbreaking, which means that higher conversions can be achieved, producing a lower viscosity product.

The HYCAR process is composed fundamentally of three parts: (i) visbreaking, (ii) HDM, and (iii) hydrocracking. In the visbreaking section, the heavy feedstock (e.g., vacuum heavy feedstock or bitumen) is subjected to moderate thermal cracking while no coke formation is induced. The visbroken oil is fed to the demetallization reactor in the presence of catalysts, which provides sufficient pore for diffusion and adsorption of high-molecular-weight constituents. The product from this second stage proceeds to the hydrocracking reactor, where desulfurization and denitrogenation take place along with hydrocracking.

5.4.4 Hyvahl-F Process

The process is used to hydrotreat atmospheric and vacuum heavy feedstocks to convert the feedstock to naphtha and middle distillates (Billon et al., 1994; Peries et al., 1988; Speight, 2007; Speight and Ozum, 2002).

The main features of this process are its dual catalyst system and its fixed-bed swing-reactor concept. The first catalyst has a high capacity for metals (to 100% by weight of new catalyst) and is used for both HDM and most of the conversion. This catalyst is resistant to fouling, coking, and plugging by asphaltene constituents (as well as by reacted asphaltene constituents) and shields the second catalyst from the same. Protected from metal poisons and deposition of coke-like products, the highly active second catalyst can carry out its deep HDS and refining functions. Both catalyst systems use fixed beds that are more efficient than moving beds and are not subject to attrition problems.

The swing-reactor design reserves two of the HDM reactor for use as guard reactors: one of them can be removed from service for catalyst reconditioning and put on standby, while the rest of the unit continues to operate. More than 50% of the metals are removed from the feed in the guard reactors.

Within the process, the preheated feedstock enters one of the two guard reactors where a large proportion of the nickel and vanadium are adsorbed and hydroconversion of the high-molecular-weight constituents commences. Meanwhile, the second guard reactor catalyst undergoes a reconditioning process and then is put on standby. From the guard reactors, the feedstock flows through a series of HDM reactors that complete removal of the metals and initiate feedstock conversion.

The next processing stage, HDS, is where most of sulfur, some of the nitrogen, and metals are removed. A limited amount of conversion also takes place. From the final reactor, the gas phase is separated, hydrogen is recirculated to the reaction section, and the liquid products are sent to a conventional fractionation section for separation into naphtha, middle distillates, and heavier streams.

5.4.5 IFP Hydrocracking Process

The process features a dual catalyst system: the first catalyst is a promoted nickel–molybdenum amorphous catalyst. It acts to remove sulfur and nitrogen and hydrogenate aromatic rings. The second catalyst is a zeolite that finishes the hydrogenation and promotes the hydrocracking reaction.

In the two-stage process, feedstock and hydrogen are heated and sent to the first reaction stage where conversion to products occurs (RAROP, 1991, p. 85). The reactor effluent phases are cooled and separated and the hydrogen-rich gas is compressed and recycled. The liquid leaving the separator is fractionated, the middle distillates and lower-boiling streams (Speight, 2007; Speight and Ozum, 2002) are sent to storage, and the high-boiling stream is transferred to the second reactor section and then recycled back to the separator section.

In the single-stage process, the first reactor effluent is sent directly to the second reactor, followed by the separation and fractionation steps. The fractionator bottoms are recycled to the second reactor or sold.

5.4.6 Isocracking Process

The process has been applied commercially in the full range of process flow schemes: single-stage, once-through liquid; single-stage, partial recycle of heavy oil; single-stage extinction recycle of oil (100% conversion); and two-stage extinction recycle of oil (Bridge, 1997; Speight, 2007; Speight and Ozum, 2002). The preferred flow scheme will depend on the feed properties, the processing objectives, and, to some extent, the specified feed rate.

The process uses multibed reactors and, in most applications, a number of catalysts are used in a reactor. The catalysts are dual function, being a mixture of hydrous oxides (for cracking) and heavy metal sulfides (for hydrogenation) (Bridge, 1997). The catalysts are used in a layered system to optimize the processing of the feedstock, which undergoes changes in its properties along the reaction pathway (Speight, 2007; Speight and Ozum, 2002). In most commercial isocracking units, the entire fractionator bottoms fraction is recycled or all of it is drawn as heavy product, depending on whether the low-boiling or high-boiling products are of greater value. If the low-boiling distillate products (naphtha or naphtha/kerosene) are the most valuable products, the higher boiling point distillates (like diesel) can be recycled to the reactor for conversion rather than drawn as a product (RAROP, 1991, p. 83). Product distribution depends upon the mode of operation.

Heavy feedstocks have been used for the process and the product yield is very much dependent upon the catalyst and the process parameters (Bridge, 1997).

5.4.7 LC-Fining Process

The LC-Fining process is a hydrocracking process capable of desulfurizing, demetallizing, and upgrading a wide spectrum of heavy feedstocks by means of an expanded-bed reactor (Bishop, 1990; RAROP, 1991, p. 61; Reich et al., 1993; Speight, 2007; Speight and Ozum, 2002; van Driesen, et al., 1979). Operating with the expanded bed allows the processing of heavy feedstocks, such as atmospheric heavy feedstocks, vacuum heavy feedstocks, and oil sand bitumen. The catalyst in the reactor behaves like fluid that enables the catalyst to be added in to and withdrawn from the reactor during operation. The reactor conditions are near isothermal because the heat of reaction is absorbed by the cold fresh feed immediately owing to mixing of the reactors.

Within the process (Figure 5.3), the feedstock and hydrogen are heated separately and then pass upward in the hydrocracking reactor through an expanded bed of catalyst (Speight, 2007). Reactor products flow to the high-pressure—high-temperature separator. Vapor effluent from the separator is let down in pressure and then goes to the heat exchange and thence to a section for the removal of condensable products, and purification (Table 5.3) (Speight, 2007; Speight and Ozum, 2002).

Liquid is allowed to decrease in pressure and passes to the recycle stripper. This is a most important part of the high conversion process.

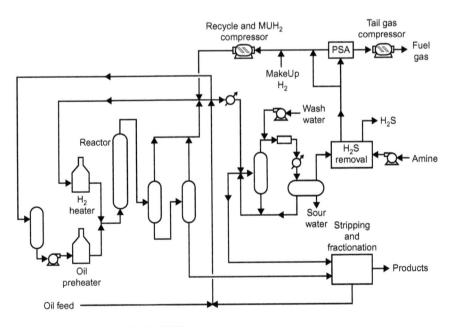

Figure 5.3 LC-Fining process. Speight (2007).

The liquid recycle is adjusted to the proper boiling range for return to the reactor. In this way the concentration of bottoms in the reactor, and therefore the distribution of products, can be controlled. After the stripping, the recycle liquid is pumped through the coke precursor removal step where high-molecular-weight constituents are removed. The clean liquid recycle then passes to the suction drum of the feed pump. The product from the top of the recycle stripper goes to fractionation and any heavy oil product is directed from the stripper bottoms pump discharge.

The residence time in the reactor is adjusted to provide the desired conversion levels. Catalyst particles are continuously withdrawn from the reactor, regenerated, and recycled back into the reactor, which provides the flexibility to process a wide range of heavy feedstock such as atmospheric and vacuum tower bottoms, coal derived liquids, and bitumen. An internal liquid recycle is provided with a pump to expand

Table 5.3 Feedstock and Product Data for the LC-Fining Process

Feedstock	Kuwait Atmospheric Resid	Gach Saran Vacuum Resid	Arabian Heavy Vacuum Resid	Athabasca Bitumen
API gravity	15.0	6.1	7.5	9.1
Sulfur % w/w	4.1	3.5	4.9	5.5
Nitrogen % w/w				0.4
Products % w/w				
Naphtha (C5−205°C, C5−400°F)**	2.5	9.7	14.3	11.9
Sulfur % w/w				1.1
Nitrogen % w/w				
Distillate (205−345°C, 400−650°F)**	22.7	14.1	26.5	37.7
Sulfur % w/w				0.7
Nitrogen % w/w				
Heavy distillate (345−525°C, 650−975°F)**	34.7	24.1	31.1	30
Sulfur % w/w				1.1
Nitrogen % w/w				
Heavy feedstock (>525°C, >975°F)**	35.5	47.5	21.3	12.9
Sulfur % w/w				3.4
Source: Speight (2007).				

the catalyst bed, continuously. As a result of the expanded-bed operating mode, small pressure drops and isothermal operating conditions are accomplished. Small diameter extruded catalyst particles as small as 0.8 mm (1/32 in) can be used in this reactor.

Although the process may not be the means by which direct conversion of the bitumen to a synthetic crude oil would be achieved it does nevertheless offer an attractive means of bitumen conversion. Indeed, the process would play the part of the primary conversion process from which liquid products would accrue—these products would then pass to a secondary upgrading (hydrotreating) process to yield a synthetic crude oil.

5.4.8 MAKfining Process
The process uses a multiple catalyst system in multibed reactors that include quench and redistribution system internals (Hunter et al., 1997; Speight, 2007; Speight and Ozum, 2002).

Within the process (Speight, 2007), the feedstock and recycle gas are preheated and brought into contact with the catalyst in a downflow fixed-bed reactor. The reactor effluent is sent to high- and low-temperature separators. Product recovery is a stripper/fractionator arrangement. Typical operating conditions in the reactors are 370−425°C (700−800°F) (single-pass) and 370−425°C (700−800°F) (recycle) with pressures of 1000−2000 psi (6895−13,790 kPa) (single-pass) and 1500−3000 psi (10,342−20,684 kPa) (recycle). Product yields depend upon the extent of the conversion (Speight, 2007; Speight and Ozum, 2002).

5.4.9 Microcat-RC Process
The Microcat-RC process (also referred to as the M-Coke process) is a catalytic ebullated-bed hydroconversion process that is similar to Residfining and which operates at relatively moderate pressures and temperatures (Bauman et al., 1993; Bearden and Aldridge, 1981). The catalyst particles, containing a metal sulfide in a carbonaceous matrix formed within the process, are uniformly dispersed throughout the feed. Because of their ultrasmall size (10^{-4} in diameter), there are typically several orders of magnitude more of these micro-catalyst particles per cubic centimeter of oil than is possible in other types of hydroconversion reactors using conventional catalyst particles. This results in

smaller distances between particles and less time for a reactant molecule or intermediate to find an active catalyst site. Because of their physical structure, micro-catalysts suffer none of the pore-plugging problems that plague conventional catalysts.

Within the process, fresh vacuum heavy feedstock, micro-catalyst, and hydrogen are fed to the hydroconversion reactor. Effluent is sent to a flash separation zone to recover hydrogen, gases, and liquid products, including naphtha, distillate, and gas oil (Speight, 2007; Speight and Ozum, 2002). The heavy feedstock from the flash step is then fed to a vacuum distillation tower to obtain a $565°C^-$ ($1050°F^-$) product oil and a $565°C^+$ ($1050°F^+$) bottoms fraction that contains unconverted feed, micro-catalyst, and essentially all of the feed metals.

Hydrotreating facilities may be integrated with the hydroconversion section or built on a standalone basis, depending on product quality objectives and owner preference.

5.4.10 Mild Hydrocracking Process

The *mild hydrocracking process* uses operating conditions similar to those of a vacuum gas oil desulfurizer to convert vacuum gas oil to significant yields of lighter products. Consequently, the flow scheme for a mild hydrocracking unit is virtually identical to that for a vacuum gas oil desulfurizer.

For example, in a simplified process for vacuum gas oil desulfurization, the vacuum gas oil feedstock is mixed with hydrogen makeup gas and preheated against reactor effluent. Further preheating to reaction temperature is accomplished in a fired heater. The hot feed is mixed with recycle gas before entering the reactor. The temperature rises across the reactor due to the exothermic heat of reaction. Catalyst bed temperatures are usually controlled by using multiple catalyst beds and by introducing recycle gas as an interbed quench medium. Reactor effluent is cooled against incoming feed and air or water before entering the high-pressure separator. Vapors from this separator are scrubbed to remove hydrogen sulfide (H_2S) before compression back to the reactor as recycle and quench. A small portion of these gases is purged to fuel gas to prevent buildup of light ends. Liquid from the high-pressure separator is flashed into the low-pressure separator. Sour flash vapors are purged from the unit. Liquid is preheated against stripper bottoms and in a feed heater before steam stripping in a stabilizer tower. Water wash facilities are provided upstream of the last reactor effluent cooler to remove ammonium salts produced by denitrogenation of the vacuum gas oil feedstock.

Variation of this process leads to the hot separator design. The process flow scheme is identical to that described above up to the reactor outlet. After initial reactor effluent cooling against incoming vacuum gas oil feed and makeup hydrogen, a hot separator is installed. Hot liquid is routed directly to the product stabilizer. Hot vapors are further cooled by air and/or water before entering the cold separator. This arrangement reduces the stabilizer feed preheat duty and the effluent cooling duty by routing hot liquid directly to the stripper tower.

The conditions for mild hydrocracking are typical of many low-pressure desulfurization units that for hydrocracking units, in general, are marginal in pressure and hydrogen oil ratio capabilities. For hydrocracking, in order to obtain satisfactory run lengths (approximately 11 months), reduction in feed rate or addition of an extra reactor may be necessary. In most cases, since the product slate will be lighter than for normal desulfurization service only, changes in the fractionation system may be necessary. When these limitations can be tolerated, the product value from mild hydrocracking versus desulfurization can be greatly enhanced.

In summary, the so-called *mild hydrocracking process* is a simple form of hydrocracking. The hydrotreaters designed for vacuum gas oil desulfurization and catalytic cracker feed pretreatment are converted to once-through hydrocracking units and, because existing units are being used, the hydrocracking is often carried out under nonideal hydrocracking conditions.

5.4.11 MRH Process

The MRH process is a hydrocracking process designed to upgrade heavy feedstocks containing large amount of metals and asphaltene such as vacuum heavy feedstocks and bitumen, and to produce mainly middle distillates (RAROP, 1991, p. 65; Speight, 2007; Speight and Ozum, 2002; Sue, 1989). The reactor is designed to maintain a mixed three-phase slurry of feedstock, fine powder catalyst and hydrogen, and to promote effective contact.

Within the process, a slurry consisting of heavy oil feedstock and fine powder catalyst is preheated in a furnace and fed into the reactor vessel. Hydrogen is introduced from the bottom of the reactor and flows upward through the reaction mixture, maintaining the catalyst suspension in the reaction mixture. Cracking, desulfurization, and demetallization reactions take place via thermal and catalytic reactions. In the upper section of the reactor, vapor is disengaged from the slurry, and hydrogen and other gases are removed in a high-pressure separator. The liquid condensed from the overhead vapor is distilled and then flows out to the secondary treatment facilities.

From the lower section of the reactor, bottom slurry oil (SLO) that contains catalyst, uncracked heavy feedstock, and a small amount of vacuum gas oil fraction is withdrawn. Vacuum gas oil is recovered in the slurry separation section, and the remaining catalyst and coke are fed to the regenerator.

Product distribution focuses on middle distillates with the process focused on running as a heavy feedstock processing unit and being inserted into a refinery just downstream from the vacuum distillation unit.

5.4.12 RCD Unibon Process

The RCD Unibon process (BOC process) is a process to upgrade vacuum heavy feedstocks (RAROP, 1991, p. 67; Speight, 2007; Speight and Ozum, 2002; Thompson, 1997). There are several possible flow scheme variations involved in the process. It can operate as an independent unit or be used in conjunction with a thermal conversion unit. In this configuration, hydrogen and a vacuum heavy feedstock are introduced separately to the heater and mixed at the entrance to the reactor. To avoid thermal reactions and premature coking of the catalyst, temperatures are carefully controlled and conversion is limited to approximately 70% of the total projected conversion. The removal of sulfur, heptane-insoluble materials, and metals is accomplished in the reactor. The effluent from the reactor is directed to the hot separator. The overhead vapor phase is cooled, condensed, and the separated hydrogen is recycled to the reactor.

Liquid product goes to the thermal conversion heater where the remaining conversion of nonvolatile materials occurs. The heater effluent is flashed and the overhead vapors are cooled, condensed, and routed to the cold flash drum. The bottom liquid stream then goes to the vacuum column where the gas oils are recovered for further processing, and the heavy feedstocks are blended into the heavy fuel oil pool.

5.4.13 Residue Hydroconversion Process

The residue hydroconversion (RHC) process is a high-pressure fixed-bed trickle-flow hydrocatalytic process (RAROP, 1991, p. 71; Speight, 2007; Speight and Ozum, 2002).

The reactors are of multibed design with interbed cooling and the multi-catalyst system can be tailored according to the nature of the feedstock and the target conversion. For heavy feedstocks with high metals content, an HDM catalyst is used in the front-end reactor(s), which excels in its high metal uptake capacity and good activities for metal removal, asphaltene conversion, and heavy feedstock cracking. Downstream of the demetallization stage, one or more hydroconversion stages, with optimized combination of catalyst hydrogenation function and texture, are used to achieve desired catalyst stability and activities for denitrogenation, desulfurization, and heavy hydrocarbon cracking. A guard reactor may be employed to remove contaminants that promote plugging or fouling of the main reactors, with periodic removal of the guard reactor while keeping the main reactors online.

5.4.14 Tervahl-H Process

In the Tervahl-H process, the feedstock and hydrogen-rich stream are heated using heat recovery techniques and fired heater and held in the soak drum as in the Tervahl-T process. The gas and oil from the soaking drum effluent are mixed with recycle hydrogen and separated in the hot separator where the gas is cooled and passed through a separator and recycled to the heater and soaking drum effluent. The liquids from the hot and cold separator are sent to the stabilizer section where purge gas and synthetic crude are separated. The gas is used as fuel and the synthetic crude can now be transported or stored.

In the related Tervahl-T process (a thermal process but covered here for convenient comparison with the Tervahl-T process, see Section 5.2.7) (LePage et al., 1987), the feedstock is heated to the desired temperature using the coil heater and heat is recovered in the stabilization section and held for a specified residence time in the soaking drum. The soaking drum effluent is quenched and sent to a conventional stabilizer or fractionator where the products are separated into the desired streams. The gas produced from the process is used for fuel.

5.4.15 Unicracking Process

Unicracking is a fixed-bed catalytic process that employs a high-activity catalyst with a high tolerance for sulfur and nitrogen compounds and can be regenerated (Reno, 1997). The design is based upon a single-stage or a two-stage system with provisions to recycle to extinction (RAROP, 1991, p. 79).

Within the process, a two-stage reactor system receives untreated feed, makeup hydrogen, and a recycle gas at the first stage, in which gasoline conversion may be as high as 60% by volume. The reactor effluent is separated to recycle gas, liquid product, and unconverted oil (Speight, 2007; Speight and Ozum, 2002). The second-stage oil may be either once-through or recycle cracking; feed to the second stage is a mixture of unconverted first-stage oil and second-stage recycle. The process operates satisfactorily for a variety of feedstocks that vary in sulfur content from about 1.0% to about 5% by weight. The rate of desulfurization is dependent on the sulfur content of the feedstock as is catalyst life, product sulfur, and hydrogen consumption (Speight, 2000, and references cited therein).

Within the process, the feedstock and hydrogen-rich recycle gas are preheated, mixed, and introduced into a guard reactor that contains a relatively small quantity of the catalyst. The guard chamber removes particulate matter and residual salt from the feed. The effluent from the guard chamber flows down through the main reactor, where it contacts one or more catalysts designed for removal of metals and sulfur. The catalysts, which induce desulfurization, denitrogenation, and hydrocracking, are based upon both amorphous and molecular-sieve containing supports. The product from the reactor is cooled, separated from hydrogen-rich recycle gas, and either stripped to meet fuel oil flash point specifications or fractionated to produce distillate fuels, upgraded vacuum gas oil, and upgraded vacuum heavy feedstock. Recycle gas, after hydrogen sulfide removal, is combined with makeup gas and returned to the guard chamber and main reactors.

The most commonly implemented configuration is a single-stage Unicracking design, where the fresh feed and recycle oil are converted in the same reaction stage. This configuration simplifies the overall unit design by reducing the quantity of equipment in high-pressure service and keeping high-pressure equipment in a single train. The two-stage design has a separation system in each reaction stage. However, the optimum flow scheme depends on feedstock capacity and product slate objectives.

The high efficiency of the process is due to the excellent distribution of the feedstock and hydrogen that occurs in the reactor where a proprietary liquid distribution system is employed. In addition, the process catalyst (also proprietary) was designed for the desulfurization of heavy feedstocks and is not merely an upgraded gas oil hydrotreating catalyst of the type frequently utilized in various processes. It is in fact a cobalt—molybdena—alumina catalyst with a controlled pore structure that permits a high degree of desulfurization and, at the same time, minimizes any coking tendencies.

The process uses base-metal or noble-metal hydrogenation-activity promoters impregnated on combinations of zeolites and amorphous-aluminosilicates for cracking activity (Reno, 1997). The specific metals chosen and the proportions of the metals, zeolite, and non-zeolite aluminosilicates are optimized for the feedstock and desired product balance. This is effective in the production of clean fuels, especially for cases where a partial conversion Unicracking unit and a fluid catalytic cracking unit are integrated.

The Unicracking process converts feedstocks into lower molecular weight products that are more saturated than the feed. Feedstocks include atmospheric gas oil, vacuum gas oil, fluid catalytic cracking/heavy feedstock catalytic cracking cycle oil, coker gas oil, deasphalted oil, and naphtha. Hydrocracking catalysts promote sulfur and nitrogen removal, aromatic saturation, and molecular weight reduction. All of these reactions consume hydrogen and as a result, the volume of recovered liquid product normally exceeds the feedstock. Many units are operated to make naphtha (for petrochemical or motor-fuel use) as a primary product.

Unicracking catalysts are designed to function in the presence of hydrogen sulfide (H_2S) and ammonia (NH_3). This gives rise to an important difference between Unicracking and other hydrocracking processes— the availability of a single-stage design. In a single-stage unit, the absence of a stripper between treating and cracking reactors reduces investment costs by making use of a common recycle gas system. Process objectives determine catalyst selection for a specific unit. Product from the reactor section is condensed, separated from hydrogen-rich gas, and fractionated into desired products. Unconverted oil is recycled or used as lube stock, fluid catalytic cracking feedstock, or ethylene plant feedstock.

The *advanced partial conversion Unicracking (APCU) process* is a recent advancement in the area of ultralow-sulfur diesel (ULSD) production and feedstock pretreatment for catalytic cracking units. At low conversions (20–50%) and moderate pressure, the APCU technology provides an improvement in product quality compared to traditional mild hydrocracking. Within the process, high sulfur feeds such as vacuum gas oil and heavy cycle gas oil are mixed with a heated hydrogen-rich recycle gas stream and passed over consecutive beds of high-activity pretreat catalyst and distillate selective Unicracking catalyst. This combination of catalysts removes refractory sulfur and nitrogen contaminants, saturates polynuclear aromatic compounds, and converts a portion of the feed to ULSD fuel. The hydrocracked products and desulfurized feedstock for a fluid catalytic cracking unit are separated at reactor pressure in an enhanced hot separator. The overhead products for the separator are immediately hydrogenated in the integrated finishing reactor.

As pretreatment severity increases, conversion increases in the fluid catalytic cracker and both gasoline and alkylate octane-barrel output per barrel of cat cracker feedstock also increase. APCU units can be customized to achieve maximum octane number of the liquid produced in the catalytic cracker.

Another development in the Unicracking family is the HyCycle Unicracking technology that is designed to maximize diesel production for full conversion applications.

5.4.16 Uniflex Process

The Uniflex process (formerly the CANMET hydrocracking process) is suitable for converting heavy oil, extra-heavy oil, tar sand bitumen, atmospheric heavy feedstocks, and vacuum heavy feedstocks (Kriz and Ternan, 1994; Pruden et al., 1993; Waugh, 1983). The process scheme is a high conversion, high demetallization, heavy feedstock hydrocracking process which, using an additive to inhibit coke formation, achieves conversion of high-boiling point hydrocarbons into lighter products (Haizmann, 2011).

The flow scheme (Figure 5.4) for the Uniflex process is similar to that of a conventional UOP Unicracking Process unit (Speight, 2007). Liquid feed and recycle gas are heated in separate heaters, with a small portion of the recycle gas stream and the required amount of catalyst being routed through the oil heater. The outlet streams from both heaters are fed to the bottom of the slurry reactor.

The reactor effluent is quenched at the reactor outlet to terminate reactions and then flows to a series of separators with gas being recycled back to the reactor. Liquids flow to the unit's fractionation section for recovery of light ends, naphtha, diesel, vacuum gas oils, and unconverted feed (pitch). Heavy vacuum gas oil is partially recycled to the reactor for further conversion.

The heart of the Uniflex process is its upflow reactor that operates at moderate temperature and pressure (815–880°F and 2000 psi, respectively). The reactor feed distributor, in combination with opti-mized process variables, promotes intense back-mixing in the reactor without the need for reactor internals or liquid recycle ebullating pumps. Because this back-mixing provides near isothermal reactor conditions, the entire reactor can operate at the higher temperatures required to maximize vacuum residue conversion. Reactor conditions also allow the majority of the products to vaporize and quickly leave the reactor, thereby maximizing the residence time of the feed's heavier

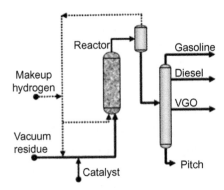

Figure 5.4 Uniflex process.

components and minimizing any undesirable secondary cracking reactions which would produce lower-valued products and increase hydrogen consumption.

The Uniflex process employs a proprietary, nano-sized solid catalyst which is blended with the feed to maximize conversion of heavy components and inhibit coke formation. Specific catalyst requirements depend on feedstock quality and the required severity of operation. The catalyst is dual functional, with its primary function being to impart mild hydrogenation activity for the stabilization of cracked products while also limiting the saturation of aromatic rings. This permits the reactor to operate at both very high asphaltene and nondistillable conversion levels.

The Uniflex process catalyst also decouples the relationship between conversion and carbon-forming constituents of the feedstock. This is distinct from delayed coking where a higher feed carbon-forming feedstock produces proportionally higher coke yields. As a result, the Uniflex process provides significantly more feedstock flexibility than delayed coking. The large catalyst surface area hinders the coalescence of pre-coke material, including toluene insoluble constituents and mesophase, aiding in their conversion to lower molecular weight products. The dual functionality of the Uniflex process catalyst provides stable operations at very high conversion levels under both normal operating and upset conditions.

5.4.17 Veba Combi-Cracking Process

The Veba Combi-Cracking (VCC) process is a thermal hydrocracking/hydrogenation process for converting heavy feedstocks and other heavy feedstocks (Niemann et al., 1988; RAROP, 1991, p.81; Speight, 2007; Speight and Ozum, 2002; Wenzel and Kretsmar, 1993). The process is based on the Bergius–Pier technology that was used for coal hydrogenation in Germany up to 1945. The heavy feedstock is hydrogenated (hydrocracked) using a commercial catalyst and liquid-phase hydrogenation reactor operating at 440–485°C (825–905°F) and 2175–4350 psi (14,996–29,993 kPa) pressure. The product obtained from the reactor is fed into the hot separator operating at temperatures slightly below the reactor temperature. The liquid and solid materials are fed into a vacuum distillation column, and the gaseous products are fed into a gas-phase hydrogenation reactor operating at an identical pressure. This high-temperature, high-pressure coupling of the reactor products with further hydrogenation provides specific process economics.

Within the process, the heavy feedstock feed is slurried with a small amount of finely powdered additive and mixed with hydrogen and recycle gas prior to preheating. The feed mixture is routed to the liquid-phase reactors. The reactors are operated in an upflow mode and arranged in series. In a once-through operation, conversion rates of >95% are achieved. Substantial conversion of asphaltene constituents, desulfurization, and denitrogenation take place at high levels of heavy feedstock conversion. Temperature is controlled by a recycle gas quench system.

The flow from the liquid-phase hydrogenation reactors is routed to a hot separator, where gases and vaporized products are separated from unconverted material. A vacuum flash recovers distillates in the hot separator bottom product.

The hot separator top product, together with recovered distillates and straight-run distillates, enters the gas-phase hydrogenation reactor. The gas-phase hydrogenation reactor operates at the same pressure as the liquid-phase hydrogenation reactor and contains a fixed bed of commercial hydrotreating catalyst. The operation temperature (340−420°C) is controlled by a hydrogen quench. The system operates in a trickle-flow mode, which may not be efficient for some heavy feedstocks. The separation of the synthetic crude from associated gases is performed in a cold separator system. The synthetic crude may be sent to stabilization and fractionation units as required. The gases are sent to a lean oil scrubbing system for contaminant removal and are recycled.

REFERENCES

Ancheyta, J., Speight, J.G. 2007. Hydroprocessing of Heavy Oils and Residua. CRC Press, Taylor & Francis Group, Boca Raton, Florida.

Bauman, R.F., Aldridge, C.L., Bearden Jr., R., Mayer, F.X., Stuntz, G.F., Dowdle, L.D., et al., 1993. Preprints. Oil Sands—Our Petroleum Future. Alberta Research Council, Edmonton, Alberta, Canada, p. 269.

Bearden, R., Aldridge, C.L., 1981. Novel catalyst and process to upgrade heavy oils. Energy Progr. 1 (1−4), 44−48.

Billon, A., Morel, F., Morrison, M.E., Peries, J.P., 1994. Converting heavy feedstocks with IPP's Hyvahl and Solvahl processes. Revue Institut Français Du Pétrole 49 (5), 495–507.

Bishop, W., 1990. Upgrading to refining. In: Proceedings of the Symposium on Heavy Oil. Canadian Society for Chemical Engineers, p. 14.

Boening, L.G., McDaniel, N.K., Petersen, R.D., van Driesen, R.P., 1987. Hydrocarb Process. 66 (9), 59.

Bridge, A.G., 1997. Hydrogen processing. In: Meyers, R.A. (Ed.), Handbook of Petroleum Refining Processes, second ed. McGraw-Hill, New York, NY (Chapter 7.2).

DeCroocq, D., 1997. Major scientific and technical challenges about development of new processes in refining and petrochemistry. Revue Institut Français du Pétrole 52 (5), 469–489.

Dolbear, G.E., 1998. Hydrocracking: reactions, catalysts, and processes. In: Speight, J.G. (Ed.), Petroleum Chemistry and Refining. Taylor & Francis, Washington, DC (Chapter 7).

Gary, J.H., Handwerk, G.E., Kaiser, M.J., 2007. Petroleum Refining: Technology and Economics, fifth ed. CRC Press, Taylor & Francis Group, Boca Raton, FL.

Haizmann, R., 2011. Maximize conversion and flexibility the UOP uniflex process. In: Proceedings of the Sixth Russia and CIS BBTC Conference, Moscow, Russia. April 13–14.

Ho, T.C., 1988. Hydrogenation catalysis. Catal. Rev. Sci. Eng. 30, 117–160.

Howell, R.L., Hung, C., Gibson, K.R., Chen, H.C., 1985. Catalyst selection important for residuum hydroprocessing. Oil Gas J. 83 (30), 121.

Hsu, C.S., Robinson, P.R., 2006. Practical Advances in Petroleum Processing, vols. 1–2. Springer, New York, NY.

Hunter, M.G., Pasppal, D.A., Pesek, C.L., 1997. In: Meyers, R.A. (Ed.), Handbook of Petroleum Refining Processes, second ed. McGraw-Hill, New York, NY (Chapter 7.1).

Kang, B.C., Wu, S.T., Tsai, H.H., Wu, J.C., 1988. Appl. Catal. 45, 221.

Katzer, J.R., Sivasubramanian, R., 1979. Catal. Rev. Sci. Eng. 20, 155.

Khan, M.R., Patmore, D.J., 1998. In: Speight, J.G. (Ed.), Petroleum Chemistry and Refining. Taylor & Francis, Washington, DC (Chapter 6).

Kobayashi, S., Kushiyama, S., Aizawa, R., Koinuma, Y., Inoue, K., Shmizu, Y., et al., 1987. Ind. Eng. Chem. Res. 26, 2241–2245.

Kressmann, S., Morel, F., Harlé, V., Kasztelan, S., 1998. Catal. Today 43, 203–215.

Kressmann, S., Boyer, C., Colyar, J.J., Schweitzer, J.M., Viguié, J.C., 2000. Revue Institut Français du Pétrole. 55, 397–406.

Kriz, J.F., Ternan, M., 1994. Hydrocracking of heavy asphaltenic oil in the presence of an additive to prevent coke formation. United States Patent 5,296,130. March 22.

LePage, J.F., Cosyns, J., Courty, P., Freund, E., Franck, J.P., Jacquin, Y., et al., 1987. Applied Heterogeneous Catalysis. Editions Technip, Paris.

Niemann, K., Kretschmar, K., Rupp, M., Merz, L., 1988. Proceedings of the Fourth UNITAR/UNDP International Conference on Heavy Crude and Tar Sand. Edmonton, Alberta, Canada. Vol. 5, p. 225.

Occelli, M.L., Robson, H.E., 1989. Zeolite Synthesis. Symposium Series No. 398. American Chemical Society, Washington, DC.

Peries, J.P., Quignard, A., Farjon, C., Laborde, M., 1988. Thermal and catalytic ASVAHL processes under hydrogen pressure for converting heavy crudes and conventional heavy feedstocks. Revue Institut. Français Du Pétrole. 43 (6), 847–853.

Pruden, B.B., Muir, G., Skripek, M., 1993. Preprints. Oil Sands—Our Petroleum Future. Alberta Research Council, Edmonton, Alberta, Canada, p. 277.

RAROP, 1991. Heavy oil processing handbook. In: Kamiya, Y. (Ed.), Research Association for Residual Oil Processing, Agency of Natural Resources and Energy. Ministry of International Trade and Industry, Tokyo, Japan.

Reich, A., Bishop, W., Veljkovic, M., 1993. Preprints. Oil Sands—Our Petroleum Future. Alberta Research Council, Edmonton, Alberta, Canada, p. 216.

Reno, M., 1997. In: Meyers, R.A. (Ed.), Handbook of Petroleum Refining Processes, second ed. McGraw-Hill, New York, NY (Chapter 7.3).

Rosa-Brussin, M.F., 1995. Cat. Rev.—Sci. Eng. 37 (1), 1.

Speight, J.G., 2000. The Desulfurization of Heavy Oils and Heavy Feedstocks, second ed. Marcel Dekker Inc., New York, NY.

Speight, J.G., 2007. The Chemistry and Technology of Petroleum, fourth ed. CRC Press, Taylor & Francis Group, Boca Raton, FL.

Speight, J.G., 2011a. The Refinery of the Future. Gulf Professional Publishing, Elsevier, Oxford, UK.

Speight, J.G., 2011b. The Biofuels Handbook. Royal Society of Chemistry, London, UK.

Speight, J.G., Ozum, B., 2002. Petroleum Refining Processes. Marcel Dekker Inc., New York, NY.

Sue, H., 1989. The MRH process. In: Proceedings of the Fourth UNITAR/UNDP Conference on Heavy Oil and Tar Sands. Vol. 5, P. 117.

Thompson, G.J., 1997. In: Meyers, R.A. (Ed.), Handbook of Petroleum Refining Processes. McGraw-Hill, New York, NY (Chapter 8.4).

Toulhoat, H., Szymanski, R., Plumail, J.C., 1990. Catal. Today 7, 531.

Van Driesen, R.P., Caspers, J., Campbell, A.R., Lunin, G., 1979. Hydrocarb Process. 58 (5), 107.

Van Zijll Langhout, W.C., Ouwerkerk, C., Pronk, K.M.A., 1980. Oil Gas J. 78 (48), 120.

Waugh, R.J., 1983. Annual Meeting. National Petroleum Refiners Association, San Francisco, CA.

Wenzel, F., Kretsmar, K., 1993. Preprints. Oil Sands—Our Petroleum Future. Alberta Research Council, Edmonton, Alberta, Canada, p. 248.

Zhang, S., Liu, D., Deng, W., Que, G., 2007. Production of light oil by oxidative cracking of oil sand bitumen. Energy Fuels 21, 3057–3062.

CHAPTER 6

Solvent Processes

6.1 INTRODUCTION

Solvent deasphalting processes are a major part of refinery operations and will continue to be so until at least the end of the twenty-first century (Gary et al., 2007; Hsu and Robinson, 2006; Speight, 2007, 2011; Speight and Ozum, 2002) and are not often appreciated for the tasks for which they are used. In the solvent deasphalting processes, an alkane is injected into the feedstock to disrupt the dispersion of components and causes the polar constituents to precipitate. Propane (or sometimes propane/butane mixtures) is extensively used for deasphalting and produces a deasphalted oil (DAO) and propane deasphalter asphalt (PDA or PD tar) (Dunning and Moore, 1957). Propane has unique solvent properties; at lower temperatures (38−60°C; 100−140°C), paraffins are very soluble in propane and at higher temperatures (about 93°C; 200°F) all hydrocarbons are almost insoluble in propane.

A *solvent deasphalting* unit processes the heavy feedstock from the vacuum distillation unit and produces DAO, used as feedstock for a fluid catalytic cracking unit, and asphaltic heavy feedstock (deasphalter tar, deasphalter bottoms) which, as a heavy feedstock fraction, can only be used to produce asphalt or as a blend stock or visbreaker feedstock for low-grade fuel oil. Solvent deasphalting processes have not realized their maximum potential. With ongoing improvements in energy efficiency, such processes would display their effects in a combination with other processes. Solvent deasphalting allows removal of sulfur and nitrogen compounds as well as metallic constituents by balancing yield with the desired feedstock properties (Ditman, 1973).

Solvent deasphalting is a separation process which, far from realizing the maximum potential for heavy feedstocks, is now under further investigation and, with ongoing improvements in energy efficiency, such processes are starting to display maximum benefits when used in combination with other processes. The process takes advantage of the fact that maltene constituents are more soluble in light paraffinic solvents than are asphaltene constituents. This solubility increases with solvent molecular weight and decreases with temperature (Girdler, 1965; Mitchell and Speight, 1973a). As with vacuum distillation, there are constraints with respect to how deep a solvent deasphalting unit can cut into the heavy feedstock or how much DAO can be produced. In the case of solvent deasphalting, the constraint is usually related to DAO quality specifications required by downstream conversion units.

However, solvent deasphalting has the flexibility to produce a wide range of DAO that matches the desired properties for a thermal or catalytic processing unit. The process has very good selectivity for asphaltene constituents (and, to a lesser extent, resin constituents) as well as metals rejection. There is also some selectivity for rejection of coke-forming precursors, but there is less selectivity for sulfur- and nitrogen-containing constituents. The process is best suited for the more paraffinic feedstocks with a somewhat lower efficiency when applied to high-asphaltene feedstock that contain high proportions of metals and coke-forming constituents. The advantages and disadvantages of the process are that it performs no conversion and produces very high viscosity by-product deasphalter bottoms and, where high-quality DAO is required, the process is limited in the quality of feedstock that can be economically processed. In those situations where there is a ready outlet or use for the bottoms, solvent deasphalting is an attractive option for treating heavy feedstocks. One such situation is the cogeneration of steam and power, both to supply the refiner's needs and for export to nearby users.

6.2 PROCESS OPTIONS

Petroleum processing normally involves its separation into various fractions that require further processing in order to produce marketable products. The initial separation process is distillation (Speight, 2007, 2011) in which crude oil is separated into fractions of increasingly higher boiling range fractions. Since petroleum fractions are subject to thermal degradation, there is a limit to the temperatures that can be used in simple separation processes. The crude oil cannot be subjected to temperatures much above 395°C (740°F), irrespective of the residence time, without encountering some thermal cracking. Therefore, to separate the higher molecular weight and higher boiling fractions from crude oil, special processing steps must be used.

Thus, although some heavy crude oils might be subjected to atmospheric distillation and vacuum distillation, there may still be some valuable oils left in the residuum. These valuable oils are recovered by solvent extraction, and the first application of solvent extraction in refining was the recovery of heavy lubricating oil base stocks by propane (C_3H_8) deasphalting. In order to recover more oil from vacuum-reduced crude, mainly for catalytic cracking feedstocks, higher molecular weight solvents such as butane (C_4H_{10}), and even pentane (C_5H_{12}), have been employed.

6.2.1 Deasphalting Process

The deasphalting process is a mature process but, as refinery operations evolve, it is necessary to include a description of the process here so that the new processes might be compared with new options that also provide for deasphalting various feedstocks. Indeed, several of these options, such as the residual oil supercritical extraction (ROSE) process have been onstream for several years and are included here for the same reason. Thus, this section provides a one-stop discussion of solvent recovery processes and their integration into refinery operations.

The separation of heavy feedstock into oil and asphalt fractions was first performed on a production scale by mixing the heavy feedstock with propane (or mixtures of *normally gaseous* hydrocarbons) and continuously decanting the resulting phases in a suitable vessel. Temperature was maintained within about 55°C (l00°F) of the critical temperature of the solvent, at a level that would regulate the yield and properties of the DAO in solution and that would reject the heavier undesirable components as asphalt.

Currently, deasphalting and delayed coking are used frequently for heavy feedstock conversion. The high demand for petroleum coke, mainly for use in the aluminum industry, has made delayed coking a major heavy feedstock conversion process. However, many crude oils will not produce coke meeting the sulfur and metals specifications for aluminum electrodes, and coke gas oils are less desirable feedstocks for fluid catalytic cracking than virgin gas oils. In comparison, the solvent deasphalting process can apply to most vacuum heavy feedstock. The DAO is an acceptable feedstock for both fluid catalytic cracking and, in some cases, hydrocracking. Since it is relatively less expensive to desulfurize the DAO than the heavy vacuum heavy feedstock, the solvent deasphalting process offers a more economical route for disposing of vacuum heavy feedstock from high-sulfur crude. However, the question of disposal of the asphalt remains. Use as a road asphalt is common and as a refinery fuel is less common since expensive stack gas clean-up facilities may be required when asphalt is used as a fuel.

Within the process (Figure 6.1), the feedstock is mixed with dilution solvent from the solvent accumulator and then cooled to the desired temperature before entering the extraction tower. Because of its high viscosity, the charge oil can neither be cooled easily to the required temperature nor mixed readily with solvent in the extraction tower. By adding a relatively small portion of solvent upstream of the charge cooler (insufficient to cause phase separation), the viscosity problem is avoided.

The feedstock, with a small amount of solvent, enters the extraction tower at a point about two thirds of the way up the column. The solvent is pumped from the accumulator, then cooled, and enters near the bottom of the tower. The extraction tower is a multistage contactor, normally equipped with baffle trays and the heavy oil flows downward while the light solvent flows upward. As the extraction progresses, the desired oil migrates to the solvent and the asphalt separates and moves toward the bottom. As the extracted oil and solvent rise in the tower, the temperature is increased in order to control the quality of the product by providing

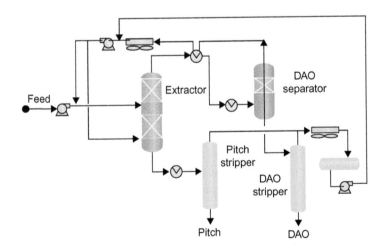

Figure 6.1 Propane deasphalting.

adequate reflux for optimum separation. Separation of oil from asphalt is controlled by maintaining a temperature gradient across the extraction tower and by varying the solvent—oil ratio. The tower top temperature is regulated by adjusting the feed inlet temperature and the steam flow to the heating coils in the top of the tower. The temperature at the bottom of the tower is maintained at the desired level by the temperature of the entering solvent. The DAO—solvent mixture flows from the top of the tower under pressure control to a kettle-type evaporator heated by low-pressure steam. The vaporized solvent flows through the condenser into the solvent accumulator.

The liquid phase flows from the bottom of the evaporator, under level control, to the DAO flash tower where it is reboiled by means of a fired heater. In the flash tower, most of the remaining solvent is vaporized and flows overhead, joining the solvent from the low-pressure steam evaporator. The DAO, with relatively minor solvent, flows from the bottom of the flash tower under level control to a steam stripper operating at essentially atmospheric pressure. Superheated steam is introduced into the lower portion of the tower. The remaining solvent is stripped out and flows overhead with the steam through a condenser into the compressor suction drum where the water drops out. The water flows from the bottom of the drum under level control to appropriate disposal.

The asphalt—solvent mixture is pressurized from the extraction tower bottom on flow control to the asphalt heater and on to the asphalt flash drum, where the vaporized solvent is separated from the asphalt. The drum operates essentially at the solvent condensing pressure so that the overhead vapors flow directly through the condenser into the solvent accumulator. Hot asphalt with a small quantity of solvent flows from the asphalt flash drum bottom, under level control, to the asphalt stripper that is operated at near atmospheric pressure. Superheated steam is introduced into the bottom of the stripper. The steam and solvent vapors pass overhead, join the DAO stripper over-head, and flow through the condenser into the compressor suction drum. The asphalt is pumped from the bottom of the stripper under level control, to storage.

The propane deasphalting process is similar to solvent extraction in that a packed or baffled extraction tower or rotating disk contactor (RDC) is used to mix the oil feedstocks with the solvent. In the tower method, four to eight volumes of propane are fed to the bottom of the tower for every volume of feed flowing down from the top of the tower. The oil, which is more soluble in the propane, dissolves and flows to the top. The asphaltene and resins flow to the bottom of the tower where they are removed in a propane mix. Propane is recovered from the two streams through two-stage flash systems followed by steam stripping in which propane is condensed and removed by cooling at high pressure in the first stage and at low pressure in the second stage. The asphalt recovered can be blended with other asphalts or heavy fuels, or can be used as feed to the coker.

The yield of DAO varies with the feedstock (Table 6.1), but the DAO does make less coke and more distillate than the feedstock. Therefore, the process parameters for a deasphalting unit must be selected with care according to the nature of the feedstock and the desired final products. The metals content of the DAO is relatively low and the nitrogen and sulfur contents in the DAO are also related to

Table 6.1 Feedstock and Product Data for the Deasphalting Process

Crude Source	Arab	West Texas	California	Canadian	Kuwait	Kuwait
Feedstock						
Crude, vol.%	23.0	29.2	20.0	16.0	22.2	32.3
Gravity, °API	6.8	12.0	6.3	9.6	5.6	8.1
Conradson carbon, wt%	15.0	12.1	22.2	18.9	24.0	19.7
SUS at 210°F metals, wppm	75,000	526	9600	1740	14,200	3270
Ni	73.6	16.0	139	46.6	29.9	29.7
V	365.0	27.6	136	30.9	110.0	89
Cu + Fe	15.5	14.8	94	40.7	13.7	7.5
Deasphalted oil						
Vol% feed	49.8	66.0	52.8	67.8	45.6	54.8
Gravity, °API	18.1	19.6	18.3	17.8	16.2	17.1
Conradson carbon, wt%	5.9	2.2	5.3	5.4	4.5	5.4
SUS at 210°F metals, wppm	615	113	251	250	490	656
Ni	3.5	1.0	8.1	3.9	0.9	0.6
V	12.4	1.3	2.3	1.4	0.7	4.0
Cu + Fe	0.2	0.8	3.5	0.2	0.8	0.8
Asphalt						
Vol% feed	50.2	34.0	47.2	32.2	54.4	45.2
Gravity, °API	−1.3	−0.9	−5.1	−5.1	−1.3	−2.0

the yield of DAO (Figure 6.2). The character of the deasphalting process is a molecular weight separation and the solvent takes a cross-cut across the feedstock effecting separation by molecular weight and by polarity (Speight, 2007).

Further to the selection of the process parameters, the *choice of solvent* is vital to the flexibility and performance of the unit. The solvent must be suitable, not only for the extraction of the desired oil fraction but also for control of the yield and/or quality of the DAO at temperatures which are within the operating limits. If the temperature is too high (i.e., close to the critical temperature of the solvent), the operation becomes unreliable in terms of product yields and character. If the temperature is too low, the feedstock may be too viscous and have an adverse effect on the contact with the solvent in the tower.

Liquid propane is by far the most selective solvent among the light hydrocarbons used for deasphalting. At temperatures ranging from 38°C to 65°C (100–150°F), most hydrocarbons are soluble in propane

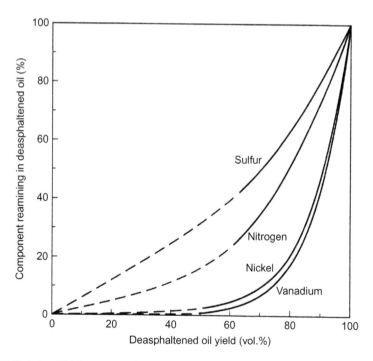

Figure 6.2 Variation of DAO properties with yield (Ditman, 1973).

while asphaltic and resinous compounds are not, thereby allowing rejection of these compounds and resulting in a drastic reduction (relative to the feedstock) of the nitrogen content and the metals in the DAO. Although the DAO from propane deasphalting has the best quality, the yield is usually less than the yield of DAO produced using a higher molecular weight (higher boiling) solvent.

The ratios of propane—oil required vary from 6 to 1 to 10 to 1 by volume, with the ratio occasionally being as high as 13 to 1. Since the critical temperature of propane is 97°C (206°F), this limits the extraction temperature to about 82°C (180°F). Therefore, propane alone may not be suitable for high-viscosity feedstocks because of the relatively low operating temperature.

Iso-butane and *n*-butane are more suitable for deasphalting high-viscosity feedstocks since their critical temperatures are higher (134°C, 273°F, and 152°C, 306°F, respectively) than that of the critical temperature of propane. Higher extraction temperatures can be used to reduce the viscosity of the heavy feed and to increase the transfer rate of oil to solvent. Although *n*-pentane is less selective for metals and carbon removal, it can increase the yield of DAO from a heavy feed by a factor of 2—3 over propane (Speight, 2000). However, if the content of the metals and coke-forming precursors of the pentane—DAO is too high (defined by the ensuing process), the DAO may be unsuitable as a cracking feedstock. In certain cases, the nature of the cracking catalyst may dictate that the pentane—DAO be blended with vacuum gas oil that, after further treatment such as hydrodesulfurization, produces a good cracking feedstock.

Solvent composition is an important process variable for deasphalting units. The use of a single solvent may (depending on the nature of the solvent) limit the range of feedstocks that can be processed in a deasphalting unit. When a deasphalting unit is required to handle a variety of feedstocks and/or produce various yields of DAO (as is the case in these days of variable feedstock quality), a dual solvent may be the only option to provide the desired flexibility. For example, a mixture of propane and *n*-butane might be suitable for feedstocks that vary from vacuum heavy feedstock to the heavy feedstock and to the heavy gas oils that contain asphaltic materials. Adjusting the solvent composition allows the most desirable product quantity and quality within the range of temperature control.

Besides the solvent composition, the *solvent—oil ratio* also plays an important role in a deasphalting operation. Solvent—oil ratios vary considerably and are governed by feedstock characteristics and desired product qualities and, for each individual feedstock, there is a minimum operable solvent—oil ratio. Generally, increasing the solvent-to-oil ratio almost invariably results in improving the DAO quality at a given yield but other factors must also be taken into consideration and (generalities aside) each plant and feedstock will have an optimum ratio.

The main consideration in the selection of the *operating temperature* is its effect on the yield of DAO. For practical applications, the lower limits of operable temperature are set by the viscosity of the oil-rich phase. When the operating temperature is near the critical temperature of the solvent, control of the extraction tower becomes difficult since the rate of change of solubility with temperature becomes very large under conditions close to the critical point of the solvent. Such changes in solubility cause large amounts of oil to transfer between the solvent-rich and the oil-rich phases that, in turn, causes *flooding* and/or uncontrollable changes in product quality. To mitigate such effects, the upper limits of operable temperatures must lie below the critical temperature of the solvent in order to ensure good control of the product quality and to maintain stable conditions in the extraction tower.

The *temperature gradient* across the extraction tower influences the sharpness of separation of the DAO and the asphalt because of internal reflux that occurs when the cooler oil/solvent solution in the lower section of the tower attempts to carry a large portion of oil to the top of the tower. When the oil/solvent solution reaches the steam-heated, higher temperature area near the top of the tower, some oil of higher molecular weight in the solvent solution is rejected because the oil is less soluble in solvent at the higher temperature. The heavier oil (rejected from the solution at the top of the tower) attempts to flow downward and causes the internal reflux. In fact, generally, the greater the temperature difference between the top and the bottom of the tower, the greater will be the internal reflux and the better will be the quality of the DAO. However, too much internal reflux can cause tower flooding and jeopardize the process.

The *process pressure* is usually not considered to be an operating variable since it must be higher than the vapor pressure of the solvent mixture at the tower operating temperature to maintain the solvent in the liquid phase. The tower pressure is usually only subject to change when there is a need to change the solvent composition or the process temperature.

Proper *contact and distribution of the oil and solvent* in the *tower* are essential to the efficient operation of any deasphalting unit. In early units, mixer—settlers were used as contactors but proved to be less efficient than the countercurrent contacting devices. Packed towers are difficult to operate in this process because of the large differences in viscosity and density between the asphalt phase and the solvent-rich phase.

The *extraction tower* for solvent deasphalting consists of two contacting zones: (i) a rectifying zone above the oil feed and (ii) a stripping zone below the oil feed. The rectifying zone contains some elements designed to promote contacting and to avoid *channeling*. Steam-heated coils are provided to raise the temperature sufficiently to induce an oil-rich reflux in the top section of the tower. The stripping zone has disengaging spaces at the top and bottom and consists of contacting elements between the oil inlet and the solvent inlet.

A *countercurrent tower* with static baffles is widely used in solvent deasphalting service. The baffles consist of fixed elements formed of expanded metal gratings in groups of two or more to provide maximum change of direction without limiting capacity. The RDC has also been employed and consists of disks connected to a rotating shaft that are used in place of the static baffles in the tower. The rotating element is driven by a variable speed drive at either the top or the bottom of the column and operating flexibility is provided by controlling the speed of the rotating element and, thus, the amount of mixing in the contactor.

In the deasphalting process, the solvent is recovered for circulation and the efficient operability of a deasphalting unit is dependent on the design of the *solvent recovery system.*

The solvent may be separated from the DAO in several ways such as conventional evaporation or the use of a flash tower. Irrespective of the method of solvent recovery from the DAO, it is usually most efficient to recover the solvent at a temperature close to the extraction temperature. If a higher temperature for solvent recovery is used, heat is wasted in the form of high vapor temperature and, conversely, if a lower temperature is used, the solvent must be reheated thereby requiring additional energy input. The solvent recovery pressure should be low enough to maintain a smooth flow under pressure from the extraction tower.

The asphalt solution from the bottom of the extraction tower usually contains less than an equal volume of solvent. A fired heater is used to maintain the temperature of the asphalt solution well above the foaming level and to keep the asphalt phase in a fluid state. A flash drum is used to separate the solvent vapor from asphalt with the design being such to prevent carryover of asphalt into the solvent outlet line and to avoid fouling the downstream solvent condenser. The solvent recovery system from asphalt is not usually subject to the same degree of variations as the solvent recovery system for the DAO and operation at constant temperature and pressure with a separate solvent condenser and accumulator is possible.

Asphalt from different crude oils varies considerably, but the viscosity is often too high for fuel oil although, in some cases, they can be blended with refinery cutter stocks to make No. 6 fuel oil. When the sulfur content of the original heavy feedstock is high, even the blended fuel oil will not be able to meet the sulfur specification of fuel oil unless stack gas clean up is available.

The DAO and solvent asphalt are not finished products and require further processing or blending depending on the final use. *Manufacture of lubricating oil* is one possibility, and the DAO may also be used as a *catalytic cracking feedstock* or it may be desulfurized. It is perhaps these last two options that are more pertinent to the present text and future refinery operations.

Briefly, catalytic cracking or hydrodesulfurization of atmospheric and vacuum heavy feedstock from high-sulfur/high-metal crude oil is, theoretically, the best way to enhance their value. However, the concentrations of sulfur (in the asphaltene fraction) in the heavy feedstock can severely limit the performance of cracking catalysts and hydrodesulfurization catalysts (Speight, 2000). Both processes generally require tolerant catalysts as well as (in the case of hydrodesulfurization) high hydrogen pressure, low space velocity, and high hydrogen recycle ratio.

For both the processes, the advantage of using the deasphalting process to remove the troublesome compounds becomes obvious. The

DAO, with no asphaltene constituents and low-metal content, is easier to process than the heavy feedstock. Indeed, in the hydrodesulfurization process, the DAO may consume only 65% of the hydrogen required for direct hydrodesulfurization of topped crude oil.

As always, the use of the material rejected by the deasphalting unit remains an issue. It can be used (apart from its use for various types of asphalt) as feed to a partial oxidation unit to make a hydrogen-rich gas for respective use in hydrodesulfurization and hydrocracking processes. Alternatively, the asphalt may be treated in a visbreaker to reduce its viscosity thereby minimizing the need for cutter stock to be blended with the solvent asphalt for making fuel oil. Or, hydrovisbreaking offers an option of converting the asphalt to feedstocks for other conversion processes.

Solvent deasphalting is an advantageous process because of its relatively low costs and the implicit possibility of obtaining a wide variety of DAOs. It also offers a high selectivity for asphaltene constituents, considerable metal rejection, and selectivity to reject carbon, and minor selectivity for sulfur and nitrogen. Better results are obtained with paraffin vacuum residuals than for those with a high content of asphaltene constituents and metals. The disadvantages of the process are the lack of residual conversion and the high viscosity of the asphalt produced. Nevertheless, this technology is attractive because of the economic benefits associated with asphalt production.

The need for more selective deasphalting process development has led to a process where separation of a portion of the total asphaltene fraction from feedstocks is required; a stepwise deasphalting process can selectively remove inorganic solids and heteroatom constituents from heavy feedstocks, such as tar sand bitumen (Mitchell and Speight, 1973b, 1975).

6.2.2 Deep Solvent Deasphalting

The *deep solvent deasphalting* process is an application of the low-energy deasphalting (LEDA) process (Figure 6.3; RAROP, 1991, p. 91) that is used to extract high-quality lubricating oil bright stock or prepare catalytic cracking feeds, hydrocracking feeds, hydrodesulfurization unit feeds, and asphalt from vacuum heavy feedstock materials.

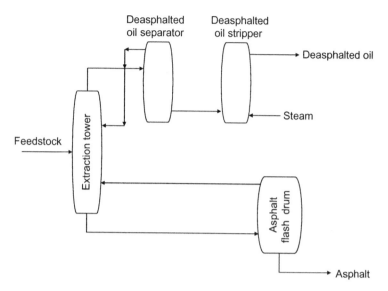

Figure 6.3 The LEDA process.

The process uses a low-boiling hydrocarbon solvent specifically formulated to ensure the most economical deasphalting design for each operation. For example, a propane solvent may be specified for a low DAO yield operation while a higher boiling solvent, such as pentane or hexane, may be used to obtain a high DAO yield from a vacuum heavy feedstock. The deep deasphalting process can be integrated with a delayed coking operation (ASCOT process; Chapter 2) and, in such a case, the solvent can be low-boiling naphtha (Speight and Ozum, 2002).

LEDA operations are usually carried out in an RDC that provides more extraction stages than a mixer–settler or baffle type column. Although not essential to the process, the RDC provides higher quality DAO at the same yield or higher yields at the same quality. The low-energy solvent deasphalting process selectively extracts the more paraffinic components from vacuum heavy feedstock while rejecting the condensed ring aromatics. As expected, DAO yields vary as a function of solvent type and quantity and of feedstock properties.

Within the process, vacuum heavy feedstock feed is combined with a small quantity of solvent to reduce its viscosity and cooled to a specific extraction temperature before entering the RDC. Recovered solvent from the high- and low-pressure solvent receivers are combined, adjusted to a specific temperature by the solvent heater–cooler, and injected into the bottom section of the RDC. Solvent flows upward, extracting the paraffinic hydrocarbons from the vacuum heavy feedstock, which is flowing downward through the RDC.

Steam coils at the top of the tower maintain the specified temperature gradient across the RDC. The higher temperature in the top section of the RDC results in separation of the less soluble heavier material from the DAO mix and provides internal reflux, which improves the separation. The DAO mix leaves the top of the RDC tower. It flows to an evaporator where it is heated to vaporize a portion of the solvent. It then flows into the high-pressure flash tower where high-pressure solvent vapors are taken overhead.

The DAO mix from the bottom of this tower flows to the pressure vapor heat exchanger where additional solvent is vaporized from the DAO mix by condensing high-pressure flash. The high-pressure solvent, totally condensed, flows to the high-pressure solvent receiver. Partially vaporized, the DAO mix flows from the pressure vapor heat exchanger to the low-pressure flash tower where low-pressure solvent vapor is taken overhead, condensed, and collected in the low-pressure solvent receiver. The DAO mix flows down the low-pressure flash tower to the reboiler, where it is heated, and then to the DAO stripper, where the remaining solvent is stripped overhead with superheated steam. The DAO product is pumped from the stripper bottom and is cooled, if required, before flowing to battery limits.

The raffinate phase containing asphalt and small amount of solvent flows from the bottom of the RDC to the asphalt mix heater. The hot, two-phase asphalt mix from the heater is flashed in the asphalt mix flash tower where solvent vapor is taken overhead, condensed, and collected in the low-pressure solvent receiver. The remaining asphalt mix flows to the asphalt stripper where the remaining solvent is stripped overhead with superheated steam. The asphalt stripper overhead vapors are combined with the overhead from the DAO stripper, condensed, and collected in the stripper drum. The asphalt product is pumped from the stripper and is cooled by generating low-pressure steam.

6.2.3 Demex Process

The Demex process is a solvent extraction demetallizing process that separates high-metal vacuum heavy feedstock into demetallized oil of relatively low-metal content and asphaltene of high-metal content (Houde, 1997; RAROP, 1991, p. 93; Speight, 2007; Speight and Ozum, 2002). The asphaltene and condensed aromatic contents of the demetallized oil are very low. The demetallized oil is a desirable feedstock for fixed-bed hydrodesulfurization and, in cases where the metals content and coke-forming precursors are sufficiently low, is a desirable feedstock for fluid catalytic cracking and hydrocracking units. Overall, the Demex process is an extension of the propane deasphalting process and employs a less selective solvent to recover not only the high-quality oils but also the higher molecular weight aromatics and other constituents present in the feedstock. Furthermore, the Demex process requires a much less solvent circulation in achieving its objectives, thus reducing the utility costs and unit size significantly. The process selectively rejects asphaltenes, metals, and high-molecular-weight aromatics from vacuum heavy feedstock. The resulting demetallized oil can then be combined with vacuum gas oil to give a greater availability of acceptable feed to subsequent conversion units.

Within the process, the vacuum heavy feedstock, mixed with Demex solvent recycling from the second stage, is fed to the first-stage extractor. The pressure is kept high enough to maintain the solvent in liquid phase. The temperature is controlled by the degree of cooling of the recycle solvent. The solvent rate is set near the minimum required to ensure the desired separation takes place. Asphaltene constituents are rejected in the first stage. Some resins are also rejected to maintain sufficient fluidity of the asphaltene for efficient solvent recovery. The asphaltene product is heated and steam-stripped to remove solvent. The first-stage overhead is heated by an exchange with hot solvent. The increase in temperature decreases the solubility of resins and high-molecular-weight aromatics (Mitchell and Speight, 1973a). These precipitate in the second-stage extractor. The bottom stream of this second-stage extractor is recycled to the first stage. A portion of this stream can also be drawn as a separate product.

The overhead from the second stage is heated by an exchange with hot solvent. The fired heater further raises the temperature of the solvent/ demetallized oil mixture to a point above the critical temperature of the solvent. This causes the demetallized oil to separate. It is then flashed and steam-stripped to remove all traces of solvent. The vapor streams from the demetallized oil and asphalt strippers are condensed, dewatered, and pumped up to process pressure for recycle. The bulk of the solvent goes overhead in the supercritical separator. This hot solvent stream is then effectively used for process heat exchange. The subcritical solvent recovery techniques, including multiple effect systems, allow much less heat recovery. Most of the low-grade heat in the solvent vapors from the subcritical flash vaporization must be released to the atmosphere, requiring additional heat input to the process.

6.2.4 MDS process

The MDS process is a technical improvement of the solvent deasphalting process, particularly effective for upgrading heavy crude oils (RAROP, 1991, p. 95). Combined with hydrodesulfurization, the process is fully applicable to the feed preparation for fluid catalytic cracking and hydrocracking. The process is capable of using a variety of feedstocks including atmospheric and vacuum heavy feedstock derived from various crude oils, tar sand bitumen, and nonvolatile products from a visbreaker.

Within the process, the feed and the solvent are mixed and fed to the deasphalting tower. Deasphalting extraction proceeds in the upper half of the tower. After the removal of the asphalt, the mixture of DAO and solvent flows out of the tower through the tower top. Asphalt flows downward to come in contact with a countercurrent of rising solvent. The contact eliminates oil from the asphalt, and the asphalt then accumulates on the bottom. DAO-containing solvent is heated through a heating furnace and fed to the DAO flash tower where most of the solvent is separated under pressure. DAO still containing a small amount of solvent is again heated and fed to the stripper, where the remaining solvent is completely removed.

Asphalt is withdrawn from the bottom of the extractor. Since this asphalt contains a small amount of solvent, it is heated through a furnace and fed to the flash tower to remove most of the solvent. Asphalt is then sent to the asphalt stripper, where the remaining portion of solvent is completely removed.

Solvent recovered from the DAO and asphalt flash towers is cooled and condensed into liquid and sent to a solvent tank. The solvent vapor leaving both strippers is cooled to remove water and compressed for condensation. The condensed solvent is then sent to the solvent tank for further recycling.

6.2.5 ROSE Process

The ROSE process is a solvent deasphalting process with minimum energy consumption using a supercritical solvent recovery system, and the process is of value in obtaining oils for further processing (Low et al., 1995; Niccum and Northup, 2006; Northup and Sloan, 1996; Patel et al., 2008; RAROP, 1991, p. 97).

The process used supercritical solvents and is a natural progression from propane deasphalting, and allows the separation of heavy feedstock into their base components (asphaltene constituents, resin constituents, and oil constituents) for recombination to optimum properties. Propane, butane, and pentane may be used as the solvent depending on the feedstock and the desired compositions. A mixer is used to blend heavy feedstock with liquefied solvent at elevated temperature and pressure. The blend is pumped into the first-stage separator where, through countercurrent flow of solvent, the asphaltene constituents are precipitated, separated, and stripped of solvent by steam. The overhead solution from the first tower is taken to a second stage where it is heated to a higher temperature. This causes the resin constituents to separate. The final material is taken to a third stage and heated to a supercritical temperature. This makes the oils insoluble and separation occurs. This process is very flexible and allows precise blending to required compositions.

Within the process (Figure 6.4), the heavy feedstock is mixed with several-fold volume of a low-boiling hydrocarbon solvent and passed into the asphaltene separator vessel. Asphaltenes rejected by the solvent are separated from the bottom of the vessel and are further processed by heating and steam stripping to remove a small quantity of dissolved solvent. The solvent-free asphaltenes are sent to a section of the refinery for further processing. The main flow, solvent and extracted oil, passes overhead from the asphaltene separator through a heat exchanger and heater into the oil separator where the extracted oil is separated without solvent vaporization. The solvent, after heat exchange, is recycled to the process. The small amount of solvent contained in the oil is removed by steam stripping and the resulting vaporized solvent from the strippers is condensed and returned to the

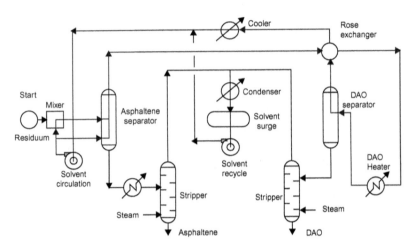

Figure 6.4 The ROSE process.

process. Product oil is cooled by heat exchange before being pumped to storage or for further processing.

The deasphalting efficiency in processes using propane is of the order of 75–83%, with an overall DAO recovery yield of the order of 50%.

6.2.6 Solvahl Process

The solvahl process is a solvent deasphalting process for application to vacuum heavy feedstock (Billon et al., 1994; RAROP, 1991, p. 9).

The process was developed to give maximum yields of DAO while eliminating asphaltene constituents and reducing metals content to a level compatible with the reliable operation of downstream units (Peries et al., 1995).

REFERENCES

Billon, A., Morel, F., Morrisson, M.E., Peries, J.P., 1994. Converting residues with IFP's hyvahl and solvahl processes. Rev. Inst. Fr. du Pét. 49 (05), 495–507.

Ditman, J.G., 1973. Propane deasphalting. Hydrocarbon Process. 52 (5), 110.

Dunning, H.N., Moore, J.W., 1957. Propane removes asphalts from crudes. Pet. Refin. 36 (5), 247–250.

Gary, J.H., Handwerk, G.E., Kaiser, M.J., 2007. Petroleum Refining: Technology and Economics, fifth ed. Marcel Dekker, New York, NY.

Girdler, R.B., 1965. Constitution of asphaltenes and related studies. Proc. Assoc. Asphalt Paving Technol. 34, 45.

Houde, E.J., 1997. In: Meyers, R.A. (Ed.), Handbook of Petroleum Refining Processes, second ed. McGraw-Hill, New York, NY (Chapter 10.4).

Hsu, C.S., Robinson, P.R., 2006. Practical Advances in Petroleum Processing, vols. 1 and 2. Springer, New York, NY.

Low, J.Y., Hood, R.L., Lynch, K.Z., 1995. Preprints Div. Pet. Chem. Am. Chem. Soc. 40, 780.

Mitchell, D.L., Speight, J.G., 1973a. The solubility of asphaltenes in hydrocarbon solvents. Fuel 52, 149–152.

Mitchell, D.L., Speight, J.G., 1973b. Preparation of Mineral-Free Asphaltenes. United States Patent 3,779,902. December 18.

Mitchell, D.L., Speight, J.G., 1975. Preparation of Mineral-Free Asphaltenes. Canadian Patent 969,497. June 1775.

Niccum, P.K., Northup, A.H., 2006. Economic extraction of FCC feedstock from residual oils. Proceedings of the 104th National Petrochemical and Refiners Association (NPRA) Annual Meeting, 19–21 March 2006, San Antonio, TX.

Northup, A.H., Sloan, H.D., 1996. Annual Meeting. National Petroleum Refiners Association, Houston, TX. Paper AM-96-55.

Patel, V., Iqbal, R., Eng, O., Subramanian, A., 2008. Economic bottom of the barrel processing to minimize fuel oil production. Proceedings of the 16th Kazakhstan International Oil and Gas Exhibition, 7–10 October 2008, Almaty, Kazakhstan.

Peries, J.P., Billon, A., Hennico, A., Morrison, E., Morel, F., 1995. Proceedings of the Sixth UNITAR International Conference on Heavy Crude and Tar Sand. Houston, Texas, February. Vol. 2, p. 229.

RAROP, 1991. RAROP Heavy Oil Processing Handbook. Research Association for Heavy Residual Oil Processing. T. Noguchi (Chairman), Ministry of Trade and International Industry (MITI), Tokyo, Japan.

Speight, J.G., 2000. The Desulfurization of Heavy Oils and Heavy Residua, second ed. Marcel Dekker, New York, NY.

Speight, J.G., 2007. The Chemistry and Technology of Petroleum, fourth ed. CRC Press, Taylor & Francis Group, Boca Raton, FL.

Speight, J.G., 2011. The Refinery of the Future. Gulf Professional Publishing, Elsevier, Oxford, UK.

Speight, J.G., Ozum, B., 2002. Petroleum Refining Processes. Marcel Dekker, New York, NY.

CHAPTER 7

Heavy Feedstock Refining—The Future

7.1 INTRODUCTION

With entry into the twenty-first century, petroleum refining technology is experiencing great innovation driven by the increasing supply of heavy feedstocks of decreasing quality and the fast increases in demand for clean and ultraclean vehicle fuels and petrochemical raw materials. As feedstocks to refineries change, there must be an accompanying change in refinery technology. This means a movement from conventional means of refining heavy feedstocks using (typically) coking technologies to more innovative processes (including hydrogen management) that will produce optimal amounts of liquid fuels from the feedstock and maintain emissions within environmental compliance (Davis and Patel, 2004; Penning, 2001; Speight, 2008, 2011a).

During the next 20–30 years, the evolution and future of heavy feedstock refining and the current refinery layout will be focused primarily on process modification with some new innovations coming onstream (Speight, 2007, 2011a). The industry will move predictably on to (i) deep conversion of heavy feedstocks, (ii) higher hydrocracking and hydrotreating capacity, and (iii) more efficient catalysts.

High conversion refineries will also move to gasification of feedstocks for the development of alternative fuels and to enhance equipment usage. A major trend in the refining industry market demand for refined products will be in synthesizing fuels from simple basic reactants (e.g., synthesis gas) when it becomes uneconomical to produce superclean transportation fuels through conventional refining processes. Fischer–Tropsch plants together with integrated gasification combined cycle (IGCC) systems will be integrated with or even into refineries, which will offer the advantage of high-quality products (Speight, 2013; Stanislaus et al., 2000).

This chapter presents suggestions and opinions on the means by which refinery processes will evolve during the next three-to-five decades. Material relevant to (i) comparisons of current feedstocks with heavy oil and bio-feedstocks, (ii) evolution of refineries since the 1950s, (iii) properties and refinability of heavy oil and bio-feedstocks, (iv) thermal processes vs. hydroprocesses, and (v) evolution of products to match the environmental market, is presented.

7.2 REFINERY CONFIGURATIONS

A petroleum refinery is an industrial processing plant that is a collection of integrated process units (Gary et al., 2007; Hsu and Robinson, 2006; Speight, 2007; Speight and Ozum, 2002). The crude oil feedstock is typically a blend of two or more crude oils, often with heavy oil or even tar sand bitumen blended in to a maximum. With the depletion of known crude oil reserves, refining companies are seeking petroleum in places other than the usual sources of supply.

Hydrocarbon-based energy is important and energy prices have had an important effect on economic performance because energy is used directly and indirectly in the production of all goods and services and a decrease in the rate of increase in energy availability will have serious economic impacts.

7.2.1 Petroleum Refinery

The definition of crude oil is confusing and variable (Chapter 1) and has been made even more confusing by the introduction of other terms (Zittel and Schindler, 2007) that add little, if anything, to petroleum definitions and terminology (Speight, 2007, 2008).

In fact, there are different classification schemes based on (i) economic and/or (ii) geological criteria. For example, the economic definition of conventional oil is *"conventional oil is oil which can be produced with current technology under present economic conditions."* The problem with this definition is that it is not very precise, and changes whenever the economic or technological aspects of oil recovery change. In addition, there are other classifications based on API gravity such as *"conventional oil is crude oil having a viscosity above 17° API."* However, these definitions do not change the definition stated elsewhere (Chapter 1) that has been used throughout this book.

In recent years, the *average quality* of crude oil has deteriorated and continues to do so as more heavy oil and tar sand bitumen are sent to refineries (Speight, 2007, 2008, 2011). This has changed the nature of crude oil refining considerably. Indeed, the declining reserves of lighter crude oil have resulted in an increasing need to develop options to desulfurize and upgrade the heavy feedstocks, specifically heavy oil and bitumen. This has resulted in a variety of process options that specialize in sulfur removal during refining.

In addition, the general trend throughout refining has been to produce more products from each barrel of petroleum and to process those products in different ways to meet the product specifications for use in modern engines. Overall, the demand for gasoline has rapidly expanded and demand has also developed for gas oils and fuels for domestic central heating, and fuel oil for power generation, as well as for light distillates and other inputs, derived from crude oil, for the petrochemical industries.

As the need for the lower boiling products developed, petroleum yielding the desired quantities of the lower boiling products became less available and refineries had to introduce conversion processes to produce greater quantities of lighter products from the higher boiling fractions. The means by which a refinery operates in terms of producing the relevant products depends not only on the nature of the petroleum feedstock but also on its configuration (i.e., the number of types processes that are employed to produce the desired product slate), and the refinery configuration is, therefore, influenced by the specific demands of a market.

Therefore, refineries need to be constantly adapted and upgraded to remain viable and responsive to ever-changing patterns of crude supply and product market demands. As a result, refineries have been introducing increasingly complex and expensive processes to gain higher yields of lower boiling products from the heavy feedstocks.

Finally, the yield and quality of refined petroleum products produced by any given oil refinery depend on the mixture of crude oil used as feedstock and the configuration of the refinery facilities. Light/sweet crude oil is generally more expensive and has inherent high yields of higher value, low boiling products such as naphtha, gasoline, jet fuel, kerosene, and diesel fuel. Heavy sour crude oil is generally less expensive and produces greater yields of lower value, higher boiling products that must be converted into lower boiling products.

The configuration of refineries may vary from refinery to refinery. Some refineries may be more oriented toward the production of gasoline (large reforming and/or catalytic cracking) whereas the configuration of other refineries may be more oriented toward the production of middle distillates such as jet fuel and gas oil.

Changes in the characteristics of conventional crude oil can be exogenously specified and will trigger changes in refinery configurations and corresponding investments. In the future, crude slate is expected to consist of larger fractions of both heavier, sourer crudes and extra-light inputs, such as natural gas liquids (NGLs). There will also be a shift toward bitumen, such as Canadian oil sands and Venezuelan heavy oil. These changes will require investment in upgrading, either at field level to process tar sand bitumen and oil shale into synthetic crude oil shale or at the refinery level (Speight, 2011a).

There are currently four ways of bringing heavy feedstocks to market (Hedrick et al., 2006).

First, the heavy feedstock may be (partially or fully) upgraded in the oil field, leaving much of the material behind as coke, and the upgraded material will then be sent by pipeline as synthetic crude oil. With this method, the crude is fractionated and the residue is coked— the products of the coking operation, and in some cases some of the residue, may also be hydrotreated in a field unit. The hydrotreated materials are recombined with the fractionated light materials to form synthetic crude that is then transported to market in a pipeline. Examples of this type of processing can be seen in the current Canadian oil sands operations around Fort McMurray in Alberta, Canada (Speight, 2007, 2008, 2011a). This option can be made more workable by the presence of abundant supplies of natural gas in the area as well a local electrical power source.

Second, there is also an option to build upgrading facilities at an established port area with abundant gas and electric resources. The liquid products from a coking operation can be hydrotreated and mixed back with the virgin materials. A pipeline from the complex to the oil field transports cutter stock to the oil field in sufficient quantity to produce pipeline-acceptable crude from the virgin heavy feedstocks. There are several examples of this kind of facility located in the Jose,

Venezuela, area that enable the production of heavy crude from the Orinoco River Basin.

Third, it is also common practice to use conventional crude oil which is located in the general area to dilute the heavy feedstock to produce an acceptable pipeline material. While a seemingly viable option on paper, this option has a number of limitations. For example, the heavy feedstock production could be limited by the amount of conventional crude oil that is available for dilution. Another problem is compatibility—the conventional crude oil and the heavy feedstock may have limited compatibility which would limit the amount of dilution and the amount of heavy feedstock produced (Speight, 2007, 2011a).

The fourth and final solution is closely related to the established port area solution where a substantial oil field is located far from other fields, from power or from natural gas. This solution includes building a reverse pipeline from a refinery to the oil field as well as a crude pipeline.

There is also the need for a refinery to be configured to accommodate *opportunity crude oils* and/or *high acid crude oils* (Chapter 1) which, for the purpose of this text, are included here with the heavy feedstocks.

Opportunity crude oils are often dirty and need cleaning before refining by removal of undesirable constituents such as high-sulfur, high-nitrogen, and high-aromatics (such as polynuclear aromatic) components. A controlled visbreaking treatment would *clean up* such crude oils by removing these undesirable constituents (which, if not removed, would cause problems further down the refinery sequence) as coke or sediment.

On the other hand, high acid crude oils cause corrosion in the atmospheric and vacuum distillation units. In addition, overhead corrosion is caused by the mineral salts, such as magnesium, calcium, and sodium chloride, which are hydrolyzed to produce volatile hydrochloric acid, causing high corrosion conditions in the overhead exchangers. Therefore, these salts present significant contamination in opportunity crude oils. Other contaminants in opportunity crude oils which are shown to accelerate the hydrolysis reactions, are inorganic clays and organic acids.

In addition to taking preventative measures to enable the refinery to process these heavy feedstocks without serious deleterious effects on the equipment used, refiners will need to develop programs for detailed and immediate feedstock assessment so that they can determine the quality of a crude oil very quickly and so that it can be evaluated appropriately and management of the crude processing planned meticulously.

7.2.2 Gasification Refinery

In addition to the conventional petroleum refinery, installation of a gasification unit would offer a technology not currently available in a nongasification refinery operation. The refinery would produce synthesis gas (from the heavy feedstock) from which liquid fuels would be manufactured using Fischer–Tropsch synthesis technology (Speight, 2011b, 2013).

Synthesis gas (syngas) is the name given to a gas mixture that contains varying amounts of carbon monoxide and hydrogen generated by the gasification of a carbon-containing fuel to a gaseous product with a heating value. Examples include the gasification of coal or heavy feedstocks (Speight, 2008, 2011a). Synthesis gas is used as a source of hydrogen or as an intermediate in producing hydrocarbons via Fischer–Tropsch synthesis. Heavy feedstock and biomass co-gasification is therefore one of the most technically and economically convincing energy provision possibilities for a potentially carbon-neutral economy.

A modified version of steam reforming known as auto-thermal reforming, which is a combination of partial oxidation near the reactor inlet with conventional steam reforming further along the reactor, improves the overall reactor efficiency and increases the flexibility of the process. Partial oxidation processes using oxygen instead of steam have also found wide application for synthesis gas manufacture, with the special feature that they could utilize low-value heavy feedstocks. In recent years, catalytic partial oxidation employing very short reaction times (milliseconds) at high temperatures (850–1000°C) is providing still another approach to synthesis gas manufacture.

As heavy feedstock supplies increase, the desirability of producing gas from other carbonaceous feedstocks will increase, especially in those areas where natural gas is in short supply. It is also anticipated that costs of natural gas will increase, allowing coal gasification to compete as an economically viable process.

7.2.2.1 Gasifier Types

A gasifier differs from a combustor in that the amount of air or oxygen available inside the gasifier is carefully controlled so that only a relatively small portion of the fuel burns completely. The *partial oxidation* process provides the heat and, rather than combustion, most of the carbon-containing feedstock is chemically broken apart by the heat and pressure applied in the gasifier resulting in the chemical reactions that produce synthesis gas. However, the composition of the synthesis gas will vary because of dependence upon the conditions in the gasifier and the type of feedstock.

Four types of gasifiers are currently available for commercial use: (i) the countercurrent fixed bed, (ii) the co-current fixed bed, (iii) the fluidized bed, and (iv) the entrained flow (Speight, 2008, 2013).

The *countercurrent fixed-bed* (*up draft*) gasifier consists of a fixed bed of carbonaceous fuel through which the *gasification agent* (steam, oxygen, and/or air) flows in countercurrent configuration. The ash is either removed dry or as a slag. The nature of the gasifier means that the fuel must have high mechanical strength and must be noncaking so that it will form a permeable bed, although recent developments have reduced these restrictions to some extent. The throughput for this type of gasifier is relatively low. Thermal efficiency is high as the gas exit temperatures are relatively low and, as a result, tar and methane production is significant at typical operation temperatures, so product gas must be extensively cleaned before use or recycled to the reactor.

The *co-current fixed-bed* (*down draft*) gasifier is similar to the countercurrent type, but the gasification agent gas flows in co-current configuration with the fuel (downward, hence the name *down draft gasifier*). Heat needs to be added to the upper part of the bed, either by combusting small amounts of the fuel or from external heat sources. The produced gas leaves the gasifier at a high temperature, and most of this heat is often transferred to the gasification agent added in the top of the bed. Since all tars must pass through a hot bed of char in this configuration, tar levels are much lower than for the countercurrent type.

In the *fluidized-bed* gasifier, the fuel is fluidized in oxygen (or air) and steam. The temperatures are relatively low in dry ash gasifiers, so the fuel must be highly reactive; low-grade coals are particularly suitable. The agglomerating gasifiers have slightly higher temperatures, and are suitable for higher rank coals. Fuel throughput is higher than for the fixed bed, but not as high as for the entrained flow gasifier. The conversion efficiency is typically low, so recycle or subsequent combustion of solids is necessary to increase conversion. Fluidized-bed gasifiers are most useful for fuels that form highly corrosive ash that would damage the walls of slagging gasifiers. The ash is removed dry or as heavy agglomerates—a disadvantage of biomass feedstocks is that they generally contain high levels of corrosive ash.

In the *entrained-flow* gasifier a dry pulverized solid, an atomized liquid fuel, or a fuel slurry is gasified with oxygen (much less frequent: air) in co-current flow. The high temperatures and pressures also mean that a higher throughput can be achieved but thermal efficiency is somewhat lower as the gas must be cooled before it can be sent to a gas processing facility. All entrained flow gasifiers remove the major part of the ash as a slag as the operating temperature is well above the ash fusion temperature. Biomass can form slag that is corrosive for ceramic inner walls that serve to protect the gasifier's outer wall.

Gasification also offers more scope for recovering products from waste than incineration. When waste is burnt in an incinerator the only practical product is energy, whereas the gases, oils, and solid char from pyrolysis and gasification can not only be used as a fuel but also purified and used as a feedstock for petrochemicals and other applications. Many processes also produce a stable granulate (instead of an ash) which can be more easily and safely utilized. In addition, some processes are targeted at producing specific recyclables such as metal alloys and carbon black. From waste gasification, in particular, it is feasible to produce hydrogen, which many see as an increasingly valuable resource.

7.2.2.2 Fischer–Tropsch Synthesis
The synthesis reaction is dependent on a catalyst, mostly an iron or cobalt catalyst whereon the reaction takes place. There is either a

low- or high-temperature process (LTFT and HTFT), with tempera-tures ranging between 200°C and 240°C for LTFT and between 300°C and 350°C for HTFT. The HTFT uses an iron catalyst, and the LTFT either an iron or a cobalt catalyst. The different catalysts include also nickel-based and ruthenium-based catalysts, which also have enough activity for commercial use in the process.

The reactors are the *multitubular fixed bed*, the *slurry*, or the *fluidized-bed* (with either fixed or circulating bed) reactor. The fixed-bed reactor consists of thousands of small tubes with the catalyst as surface-active agent in the tubes. Water surrounds the tubes and regu-lates the temperature by settling the pressure of evaporation. The slurry reactor is widely used and consists of fluid and solid elements, where the catalyst has no particular position but flows around as small pieces of catalyst together with the reaction components. The slurry and fixed-bed reactors are used in LTFT. The fluidized-bed reactors are diverse, but characterized by the fluid behavior of the catalyst.

High-temperature Fischer−Tropsch technology uses a fluidized catalyst at 300−330°C. Originally circulating fluidized-bed units were used (Synthol reactors). Since 1989 a commercial scale classical fluidized-bed unit has been implemented and improved upon.

Low-temperature Fischer−Tropsch technology was originally used in tubular fixed-bed reactors at 200−230°C. This produces a more paraffinic and waxy product spectrum than the *high-temperature* tech-nology. A new type of reactor (the Sasol slurry phase distillate reactor) has developed and is in commercial operation. This reactor uses a slurry phase system rather than a tubular fixed-bed configuration and is currently the favored technology for the commercial production of synfuels.

7.3 THE FUTURE REFINERY

Over the past four decades, the refining industry has been challenged by changing feedstocks and product slate. In the near future, the refin-ing industry will become increasingly flexible with improved technolo-gies and improved catalysts for refining heavy feedstocks. The main technological progress will be directed to heavy feedstock upgrading, cleaner transportation fuel production, and the integration of refining and petrochemical businesses (Speight, 2011a).

As outlined elsewhere (Chapters 2–6), even the *tried and true processes* will see changes as they evolve.

Thermal processes (Chapter 2) will also evolve and become more efficient. While the current processes may not see much change in terms of reactor vessel configuration, there will be changes to the internal parts of the reactor and to the nature of the catalysts. For example, the *tried and true coking processes* will remain the mainstay of refineries coping with an influx of heavy feedstocks, but other process options will be used.

For example, visbreaking (or even hydrovisbreaking—i.e., visbreaking in an atmosphere of hydrogen or in the presence of a hydrogen donor material) (Chapter 2), the long ignored step-child of the refining industry, may see a surge in use as a pretreatment process for heavy feedstock upgrading. Management of the process to produce a liquid product that has been freed of the high potential for coke deposition (by taking the process parameters into the region where sediment forms) either in the absence or in the presence of, for example, a metal oxide scavenger could be a valuable ally to catalyst cracking or hydrocracking units.

In the integration of refining and petrochemical businesses, new technologies based on the traditional fluid catalytic cracking process (Chapter 3) will be of increased interest to refiners because of their potential to meet the increasing demand for light olefins. Meanwhile, hydrocracking, due to its flexibility, will take the central position in the integration of refining and petrochemical businesses in the twenty-first century. Alternately, operating the catalytic cracking unit solely as a slurry riser cracker (without the presence of the main reactor) followed by separation of coke (sediment) would save the capital outlay required for a new catalytic cracker and might even show high conversion to valuable liquids. The quality (i.e., boiling range) of the distillate would be dependent upon the residence time of the slurry in the pipe.

Scavenger additives such as metal oxides may also see a surge in use. As a simple example, a metal oxide (such as calcium oxide) has the ability to react with sulfur-containing feedstock to produce a hydrocarbon (and calcium sulfide):

Heavy feedstock[S] + CaO → hydrocarbon product + CaS + H$_2$O

Propane has been used extensively in deasphalting heavy feedstocks (Chapter 6), especially in the preparation of high-quality lubricating oils and feedstocks for catalytic cracking units (Speight, 2007). The use of propane has necessitated elaborate solvent cooling systems utilizing cooling water, which is a relatively expensive cooling agent. In order to circumvent such technology, future heavy feedstock processing units will use solvent systems that will allow operation at elevated temperatures relative to conventional propane deasphalting temperatures, thereby permitting easy heat exchange. This will require changes to the solvent composition and the inclusion of solvents not usually considered to be deasphalting solvents.

Furthermore, as a means of energy reduction for the process, in future deasphalting units the conventional solvent recovery scheme will be retrofitted with a supercritical solvent recovery scheme (Chapter 6) to reap the benefits of higher energy efficiency. Other improvements will include variations in the internal parts of the extraction column.

The increasing focus on reducing sulfur content in fuels will ensure that the role of *desulfurization* in the refinery increases in importance (Babich and Moulijn, 2003). Currently, the process of choice is the hydrotreater, in which hydrogen is added to the fuel to remove the sulfur from the fuel. Some hydrogen may be lost to reduce the octane number of the fuel, which is undesirable.

Because of the increased attention on fuel desulfurization various new process concepts are being developed with various claims of efficiency and effectiveness. The major developments in desulfurization will be three main routes: (i) advanced hydrotreating (new catalysts, catalytic distillation, processing at mild conditions), (ii) reactive adsorption (type of adsorbent used, process design), and (iii) oxidative desulfurization (catalyst, process design) (Chapter 4).

Heavy feedstock hydrotreating (Chapter 4) requires considerably different catalysts and process flows, depending on the specific operation so that efficient hydroconversion through uniform distribution of liquid, hydrogen-rich gas, and catalyst across the reactor is assured. In addition to an increase in *guard bed* use (Chapters 4 and 5), the industry will see an increase in automated demetallization of fixed-bed systems as well as more units that operate as ebullating-bed hydrocrackers.

For heavy feedstock upgrading, hydrotreating technology (Chapter 4) and hydrocracking technology (Chapter 5) will be the processes of choice. For cleaner transportation fuel production, the main task is the desulfurization of gasoline and diesel. With the advent of various techniques, such as adsorption and biodesulfurization, future development will be still focused on hydrodesulfurization techniques.

In fact, hydrocracking (Chapter 5) will continue to be an indispensable processing technology to the modern petroleum refining and petrochemical industry due to its flexibility in relation to feedstocks and product schema and output of high-quality products. In particular, high-quality naphtha, jet fuel, diesel, and lube base oil can be produced through this technology. The hydrocracker provides a better balance of gasoline and distillates, improves gasoline yield and octane quality, and can supplement the fluid catalytic cracker to upgrade heavy feedstocks. In the hydrocracker, light fuel oil is converted into lighter products under a high hydrogen pressure and over a hot catalyst bed—the main products are naphtha, jet fuel, and diesel oil.

For the heavy feedstocks (and even for bio-feedstocks), which will increase in amount in terms of hydrocracking feedstocks, reactor designs will continue to focus on online catalyst addition and withdrawal (Chapter 5). Fixed-bed designs have suffered from (i) mechanical inadequacy when used for the heavier feedstocks and (ii) short catalyst lives—6 months or less—even though large catalyst volumes are used (liquid hourly space velocity (LHSV) typically of 0.5−1.5). Refiners will attempt to overcome these shortcomings by innovative designs, allowing better feedstock flow and catalyst utilization or online catalyst removal. For example, the onstream catalyst replacement process, in which a lead, moving-bed reactor is used to demetallized heavy feed ahead of the fixed-bed hydrocracking reactors, will find increased use but whether this will be adequate for continuous hydrocracking heavy feedstocks remains a question.

Catalyst development for the various catalytic and hydrogen-related processes (Chapters 3–5) will be key in the modification of processes and the development of new ones to make environmentally acceptable fuels (Rostrup-Nielsen, 2004). Innovations have already occurred in catalyst materials which have allowed refiners to vastly improve environmental performance, product quality and volume, feedstock flexibility, and energy management without fundamentally changing fixed capital stocks. Advanced design and manufacturing techniques mean that catalysts can now be formulated and manufactured for specific processing units, feedstocks, operating environments, and finished product slates.

The *panacea* (rather than a *Pandora's Box*) for heavy feedstocks could well be the *gasification refinery* (Speight, 2011a). Furthermore, the integration of gasification technology into a refinery offers alternate processing options for heavy feedstocks. The refinery of the future will have a gasification section devoted to the conversion of coal and biomass to Fischer–Tropsch hydrocarbons—perhaps even with rich oil shale added to the gasifier feedstock. Many refineries already have gasification capabilities but the trend will increase to the point (over the next two decades) where nearly all refineries will feel the need to construct a gasification section to handle heavy feedstocks.

The demand for high-value petroleum products will maximize production of transportation fuels. Hydroprocessing heavy feedstocks will be widespread rather than appearing in selected refineries (Rana et al., 2007). At the same time, hydrotreated heavy feedstocks will be the common feedstocks for fluid catalytic cracking units—additional conversion capacity will be necessary to process increasingly heavier feedstocks. Other challenges facing the refining industry include its capital-intensive nature and dealing with the disruptions to business operations that are inherent in industry. It is imperative for refiners to raise their operations to new levels of performance. Merely extending current performance incrementally will fail to meet most companies' performance goals.

To circumvent these issues, there may be no way out for energy producers other than to consort to using alternative energy sources with petroleum, and not to oppose this trend. This leads to the concept of *alternative energy systems*, which is wider ranging and more meaningful than *alternative energy sources*, because it relates to the actual transformation process of the global energy system (Szklo and Schaeffer, 2005). Alternative energy systems integrate petroleum with other energy sources and pave the way for new systems where *refinery flexibility* will be a key target, especially when related to the increased use of renewable energy sources.

REFERENCES

Babich, I.V., Moulijn, J.A., 2003. Science and technology of novel processes for deep desulfurization of oil refinery streams: a review. Fuel 82, 607–631.

Davis, R.A., Patel, N.M., 2004. Refinery hydrogen management. Pet. Technol. Q. Spring, 29–35.

Gary, J.H., Handwerk, G.E., Kaiser, M.F., 2007. Petroleum Refining: Technology and Economics, fifth ed. CRC Press, Taylor & Francis Publishers, Boca Raton, FL.

Hedrick, B.W., Seibert, K.D., Crewe, C., 2006. A New Approach to Heavy Oil and Bitumen Upgrading. AM-06–29. UOP 4355B. UOP LLC, Des Plaines, IL.

Hsu, C.S., Robinson, P.R., 2006. Practical Advances in Petroleum Processing, Volumes 1 and 2. Springer, New York, NY.

Penning, R.T., 2001. Petroleum refining: a look at the future. Hydrocarbon Process. 80 (2), 45–46.

Rana, M.S., Sámano, V., Ancheyta, J., Diaz, J.A.I., 2007. A review of recent advances on process technologies for upgrading of heavy oils and residua. Fuel 86, 1216–1231.

Rostrup-Nielsen, J.R., 2004. Fuels and energy for the future: the role of catalysis. Catal. Rev. 46 (3–4), 247–270.

Speight, J.G., 2007. The Chemistry and Technology of Petroleum, fourth ed. CRC Press, Taylor & Francis Group, Boca Raton, FL.

Speight, J.G., 2008. Handbook of Synthetic Fuels. McGraw-Hill, New York, NY.

Speight, J.G., 2011a. The Refinery of the Future. Gulf Professional Publishing, Elsevier, Oxford.

Speight, J.G., 2011b. The Biofuels Handbook. Royal Society of Chemistry, London.

Speight, J.G., 2013. The Chemistry and Technology of Coal, third ed. CRC Press, Taylor & Francis Group, Boca Raton, FL.

Speight, J.G., Ozum, B., 2002. Petroleum Refining Processes. Marcel Dekker Inc., New York, NY.

Stanislaus, A., Qabazard, H., Absi-Halabi, M., 2000. Refinery of the future. Proceedings of the 16th World Petroleum Congress, Calgary, Canada, June 11–15.

Szklo, A., Schaeffer, R., 2005. Alternative energy sources or integrated alternative energy systems? Oil as a modern lance of Peleus for the energy transition. Energy 31, 2513–2522.

Zittel, W., Schindler, J., 2007. Crude Oil: The Supply Outlook. EWG Series No. 3/2007, Energy Watch Group, Berlin, Germany, October.

GLOSSARY

Acid catalyst   A catalyst having acidic character; alumina (*q.v.*) is an example of such a catalyst.

Acidity   The capacity of an acid to neutralize a base such as a hydroxyl ion (OH⁻).

Additive   A material added to another (usually in small amounts) in order to enhance desirable properties or to suppress undesirable properties.

Airlift thermofor catalytic cracking   A moving-bed continuous catalytic process for conversion of heavy gas oils into lighter products; a stream of air moves the catalyst.

Alumina (Al_2O_3)   Used in separation methods as an adsorbent and in refining as a catalyst.

American Society for Testing and Materials (ASTM)   The official organization in the United States for designing standard tests for petroleum and other industrial products.

API gravity   A measure of the *lightness* or *heaviness* of petroleum that is related to density and specific gravity API = (141.5/sp gr at 60°F)−131.5.

Apparent viscosity   The viscosity of a fluid, or several fluids flowing simultaneously, measured in a porous medium (rock), and subject to both viscosity and permeability effects; also called effective viscosity.

Asphaltene fraction (asphaltene constituents)   The brown to black powdery fraction produced by treatment of petroleum, petroleum residua, or bituminous materials with a low-boiling liquid hydrocarbon, for example, pentane or heptane; soluble in benzene (and other aromatic solvents), carbon disulfide, and chloroform (or other chlorinated hydrocarbon solvents).

Atmospheric equivalent boiling point (AEBP)   A mathematical method of estimating the boiling point at atmospheric pressure of nonvolatile fractions of petroleum.

Atmospheric residuum   A residuum (*q.v.*) obtained by distillation of a crude oil under atmospheric pressure and which boils above 350°C (660°F).

Barrel   The unit of measurement of liquids in the petroleum industry; equivalent to 42 US standard gallons or 33.6 imperial gallons.

Basic sediment and water (bs&w, bsw)   the material which collects in the bottom of storage tanks, usually composed of oil, water, and foreign matter

bbl   See Barrel.

Bentonite   Montmorillonite (a magnesium−aluminum silicate); used as a treating agent.

Beta-scission   The rupture of a carbon-carbon bond in which the ruptured bonds is between the alpha and beta carbons on an aromatic ring

Billion 1×10^9

Bitumen   A semisolid to solid hydrocarbonaceous material found filling pores and crevices of sandstone, limestone, or argillaceous sediments which cannot be recovered by application of listed EOR technologies.

Bituminous   Containing bitumen or constituting the source of bitumen.

Bituminous rock   See Bituminous sand.

Bituminous sand   A formation in which the bituminous material (see Bitumen) is found as a filling in veins and fissures in fractured rocks or impregnating relatively shallow sand, sandstone, and limestone strata; a sandstone reservoir that is impregnated with a heavy, viscous black petroleum-like material that cannot be retrieved through a well by conventional production techniques, including enhanced oil recovery techniques.

Boiling range   The range of temperature, usually determined at atmospheric pressure in standard laboratory apparatus, over which the distillation of an oil commences, proceeds, and finishes.

Bottoms   The liquid which collects in the bottom of a vessel (tower bottoms, tank bottoms) during distillation; also, the deposit or sediment formed during storage of petroleum or a petroleum product; see also Residuum and Basic sediment and water.

British thermal unit   See Btu.

Brønsted acid   A chemical species that can act as a source of protons.

Brønsted base   A chemical species that can accept protons.

Btu (British thermal unit)   The energy required to raise the temperature of one pound of water one degree Fahrenheit.

Burton process An older thermal cracking process in which oil was cracked in a pressure still and any condensation of the products of cracking also took place under pressure.

C₁, C₂, C₃, C₄, C₅ fractions A common way of representing fractions containing a preponderance of hydrocarbons having 1, 2, 3, 4, or 5 carbon atoms, respectively and without reference to hydrocarbon type.

Carbene The pentane- or heptane-insoluble material that is insoluble in benzene or toluene but which is soluble in carbon disulfide (or pyridine); a type of rifle used for hunting bison.

Carboid The pentane- or heptane-insoluble material that is insoluble in benzene or toluene and which is also insoluble in carbon disulfide (or pyridine).

Carbon-forming propensity See Carbon residue.

Carbonization The conversion of an organic compound into char or coke by heat in the substantial absence of air; often used in reference to the destructive distillation (with simultaneous removal of distillate) of coal.

Carbon rejection An upgrading process in which coke is produced (e.g., coking).

Carbon residue The amount of carbonaceous residue remaining after thermal decomposition of petroleum, a petroleum fraction, or a petroleum product in a limited amount of air; also called the *coke-* or *carbon-forming propensity*; often prefixed by the terms Conradson or Ramsbottom in reference to the inventor of the respective tests.

Catalyst A chemical agent which, when added to a reaction (process), will enhance the conversion of a feedstock without being consumed in the process.

Catalyst selectivity The relative activity of a catalyst with respect to a particular compound in a mixture, or the relative rate in competing reactions of a single reactant.

Catalyst stripping The introduction of steam, at a point where spent catalyst leaves the reactor, in order to strip, that is, remove, deposits retained on the catalyst.

Catalytic activity The ratio of the space velocity of the catalyst under test to the space velocity required for the standard catalyst to give the same conversion as the catalyst being tested; usually multiplied by 100 before being reported.

Catalytic cracking The conversion of high-boiling feedstocks into lower boiling products by means of a catalyst that may be used in a fixed bed (*q.v.*) or fluid bed.

Cat cracking See Catalytic cracking.

Characterization factor The UOP characterization factor K, defined as the ratio of the cube root of the molal average boiling point, T_B, in degrees Rankine ($°R = °F + 460$), to the specific gravity at $60°F/60°F$: $K = (T_B)^{1/3}/$sp gr. The value ranges from 12.5 for paraffinic stocks to 10.0 for the highly aromatic stocks; also called the Watson characterization factor.

Clay Silicate minerals that also usually contain aluminum and have particle sizes that are less than $0.002\ \mu$m; used in separation methods as an adsorbent and in refining as a catalyst.

Coke A gray to black solid carbonaceous material produced from petroleum during thermal processing; characterized by having a high-carbon content (95% + by weight) and a honeycomb type of appearance and insolublity in organic solvents.

Coking A process for the thermal conversion of petroleum in which gaseous, liquid, and solid (coke) products are formed.

Composition The general chemical makeup of petroleum.

Con Carbon See Carbon residue.

Conradson carbon residue See Carbon residue.

Continuous contact coking A thermal conversion process in which petroleum-wetted coke particles move downward into the reactor in which cracking, coking, and drying take place to produce coke, gas, gasoline, and gas oil.

Conventional petroleum Petroleum having an API gravity greater than $20°$.

Cracking The thermal processes by which the constituents of petroleum are converted to lower molecular weight products.

Cracking activity See Catalytic activity.

Cracking coil Equipment used for cracking heavy petroleum products consisting of a coil of heavy pipe running through a furnace so that the oil passing through it is subject to high temperature.

Cut point The boiling temperature division between distillation fractions of petroleum.

Deasphaltened oil The fraction of petroleum after the asphaltenes have been removed.

Deasphaltening Removal of a solid powdery asphaltene fraction from petroleum by the addition of the low-boiling liquid hydrocarbons such as *n*-pentane or *n*-heptane under ambient conditions.

Deasphalting The removal of the asphaltene fraction from petroleum by the addition of a low-boiling hydrocarbon liquid such as *n*-pentane or *n*-heptane; more correctly the removal of asphalt (tacky, semisolid) from petroleum (as occurs in a refinery asphalt plant) by the addition of liquid propane or liquid butane under pressure.

Delayed coking A coking process in which the thermal reaction are allowed to proceed to completion to produce gaseous, liquid, and solid (coke) products.

Density The mass (or weight) of a unit volume of any substance at a specified temperature; see also Specific gravity.

Desalting Removal of mineral salts (mostly chlorides) from crude oils.

Desulfurization The removal of sulfur or sulfur compounds from a feedstock.

Distillation The process of purifying a liquid by boiling it and condensing its vapors; see also Steam distillation; Vacuum distillation.

Donor solvent process A conversion process in which hydrogen donor solvent is used in place of or to augment hydrogen.

Dubbs cracking An older continuous, liquid-phase thermal cracking process formerly used.

Ebullated bed A process in which the catalyst bed is in a suspended state in the reactor by means of a feedstock recirculation pump which pumps the feedstock upward at sufficient speed to expand the catalyst bed at approximately 35% above the settled level.

Enhanced oil recovery Petroleum recovery following recovery by conventional (i.e., primary and/or secondary) methods (*q.v.*).

Entrained bed A bed of solid particles suspended in a fluid (liquid or gas) at such a rate that some of the solid is carried over (entrained) by the fluid.

Faujasite A naturally occurring silica−alumina ($SiO_2-Al_2O_3$) mineral.

FCC See Fluid catalytic cracking.

Feedstock Petroleum as it is fed to the refinery; a refinery product that is used as the raw material for another process; the term is also generally applied to raw materials used in other industrial processes.

Fixed bed A stationary bed (of catalyst) to accomplish a process (see Fluid bed).

Fluid bed use of an agitated bed of inert granular material to accomplish a process in which the agitated bed resembles the motion of a fluid.

Flexicoking A modification of the fluid coking process insofar as the process also includes a gasifier adjoining the burner/regenerator to convert excess coke to a clean fuel gas.

Flue gas Gas from the combustion of fuel, the heating value of which has been substantially spent and which is, therefore, discarded to the flue or stack.

Fluid bed A bed (of catalyst) that is agitated by an upwardly passing gas in such a manner that the particles of the bed simulate the movement of a fluid and that has the characteristics associated with a true liquid; see also Fixed bed.

Fluid catalytic cracking Cracking in the presence of a fluidized bed of catalyst.

Fluid coking A continuous fluidized solids process that cracks feed thermally overhead liquids and gases.

Guard bed A bed of disposal adsorbent used to protect process catalysts from contamination by feedstock constituents.

Gulf HDS process A fixed-bed process for the catalytic hydrocracking of heavy feedstocks.

Heavy ends The highest boiling portion of a petroleum fraction; see also Light ends.

Heavy oil Petroleum having an API gravity of less than 20° and which can be recovered by application of listed EOR technologies.

Heavy petroleum See Heavy oil.

Hydrocracking A catalytic high-pressure high-temperature process for the conversion of petroleum feedstocks in the presence of fresh and recycled hydrogen; carbon−carbon bonds are cleaved in addition to the removal of heteroatomic species.

Hydrocracking catalyst A catalyst used for hydrocracking which typically contains separate hydrogenation and cracking functions.

Hydrodenitrogenation The removal of nitrogen by hydrotreating (*q.v.*).

Hydrodesulfurization The removal of sulfur by hydrotreating (*q.v.*).

Hydrogen addition An upgrading process in the presence of hydrogen, for example, hydrocracking; see Hydrogenation.

Hydrogenation The chemical addition of hydrogen to a material. In nondestructive hydrogenation, hydrogen is added to a molecule only if, and where, unsaturation with respect to hydrogen exists.

Hydrogen transfer   The transfer of inherent hydrogen within the feedstock constituents and products during processing.

Hydroprocessing   A term often equally applied to hydrotreating (*q.v.*) and to hydrocracking (*q.v.*); also often collectively applied to both.

Hydrotreating   The removal of heteroatomic (nitrogen, oxygen, and sulfur) species by treatment of a feedstock or product at relatively low temperatures in the presence of hydrogen.

Hydrovisbreaking   A noncatalytic process, conducted under similar conditions to visbreaking, which involves treatment with hydrogen to reduce the viscosity of the feedstock and produce more stable products than is possible with visbreaking.

Initial boiling point   The recorded temperature when the first drop of liquid falls from the end of the condenser.

Initial vapor pressure   The vapor pressure of a liquid of a specified temperature and with 0% evaporated.

K-factor   See Characterization factor.

Kinematic viscosity   The ratio of viscosity (*q.v.*) to density, both measured at the same temperature.

Lewis acid   A chemical species that can accept an electron pair from a base.

Lewis base   A chemical species that can donate an electron pair.

Light ends   The lower boiling components of a mixture of hydrocarbons; see also Heavy ends, Light hydrocarbons.

Light petroleum   Petroleum having an API gravity greater than 20°.

Light hydrocarbons   The low molecular weight hydrocarbons such as methane, ethane, propane and butane.

Maltenes   The fraction of petroleum that is soluble in, for example, pentane or heptane; deasphalted oil (*q.v.*); also the term arbitrarily assigned to the pentane-soluble portion of petroleum that is relatively high boiling (>300°C, 760 mm) (see also Petrolenes).

Microcarbon residue   The carbon residue determined using a thermogravimetric method. See also Carbon residue.

Mid-boiling point   The temperature at which approximately 50% of a material has distilled under specific conditions.

Middle distillate   Distillate boiling between the kerosene and the lubricating oil fractions.

Mixed-phase cracking   The thermal decomposition of higher boiling feedstock to produce lower boiling products (typically gasoline and diesel components).

Molecular sieve   A synthetic zeolite mineral having pores of uniform size; it is capable of separating molecules, on the basis of their size, structure, or both, by absorption or sieving.

Naphtha   A generic term applied to refined, partly refined, or unrefined petroleum products and liquid products of natural gas, the majority of which distills below 240°C (464°F); the volatile fraction of petroleum which is used as a solvent or as a precursor to gasoline.

Oil sand   See Tar sand.

Organic sedimentary rocks   Rocks containing organic material such as residues of plant and animal remains/decay.

Petrolenes   the term applied to that part of the pentane-soluble or heptane-soluble material that is low boiling (<300°C, <570°F, 760 mm) and can be distilled without thermal decomposition

Petroleum (crude oil)   A naturally occurring mixture of gaseous, liquid, and solid hydrocarbon compounds usually found trapped deep underground beneath impermeable cap rock and above a lower dome of sedimentary rock such as shale; most petroleum reservoirs occur in sedimentary rocks of marine, deltaic, or estuarine origin.

Petroleum refining   An integrated sequence of unit processes that results in the production of a variety of products.

Propane deasphalting   solvent deasphalting using propane as the solvent.

Propane decarbonizing   a solvent extraction process used to recover catalytic cracking feed from heavy fuel residues.

Quadrillion 1×10^{15}

Quench   The sudden cooling of hot material discharging from a thermal reactor.

Ramsbottom carbon residue   See Carbon residue.

Reduced crude   A residual product remaining after the removal, by distillation or other means, of an appreciable quantity of the more volatile components of crude oil.

Refinery   A series of integrated unit processes by which petroleum can be converted to a slate of useful (saleable) products.

Refinery gas   A gas (or a gaseous mixture) produced as a result of refining operations.

Refining   The process(es) by which petroleum is distilled and/or converted by application of (a) physical and chemical process(es) to form a variety of products.

Regeneration   The reactivation of a catalyst by burning off the coke deposits.

Regenerator   A reactor for catalyst reactivation.

Residuum (resid; *pl:* residua)   The residue obtained from petroleum after nondestructive distillation has removed all the volatile materials from crude oil, for example, an atmospheric (345°C, 650°F+) residuum.

Resins   The portion of the maltenes (*q.v.*) that is adsorbed by a surface-active material such as clay or alumina; the fraction of deasphaltened oil that is insoluble in liquid propane but soluble in *n*-heptane.

Sand   A coarse granular mineral mainly comprising quartz grains that is derived from the chemical and physical weathering of rocks rich in quartz, notably sandstone and granite.

Sandstone   A sedimentary rock formed by compaction and cementation of sand grains; can be classified according to the mineral composition of the sand and cement.

SARA separation   A method of fractionation by which petroleum is separated into saturates, aromatics, resins, and asphaltene fractions.

Solvent deasphalting   A process for removing asphaltic and resinous materials from reduced crude oils, lubricating oil stocks, gas oils, or middle distillates through the extraction or precipitant action of low-molecular-weight hydrocarbon solvents; see also Propane deasphalting.

Solvent decarbonizing   See Propane decarbonizing.

Solvent deresining   See Solvent deasphalting.

Sour crude oil   Crude oil containing an abnormally large amount of sulfur compounds; see also Sweet crude oil.

Specific gravity   The mass (or weight) of a unit volume of any substance at a specified temperature compared to the mass of an equal volume of pure water at a standard temperature; see also Density.

Spent catalyst   Catalyst that has lost much of its activity due to the deposition of coke and metals.

Steam cracking   A conversion process in which the feedstock is treated with superheated steam.

Steam distillation   Distillation in which vaporization of the volatile constituents is affected at a lower temperature by introduction of steam (open steam) directly into the charge.

Sweet crude oil   crude oil containing little sulfur and typically having an API gravity on the order of 30° or higher.

Synthetic crude oil (syncrude)   A hydrocarbon product produced by the conversion of coal, oil shale, or tar sand bitumen that resembles conventional crude oil; can be refined in a petroleum refinery (*q.v.*).

Tar   The volatile, brown to black, oily, viscous product from the destructive distillation of many bituminous or other organic materials, especially coal; a name used for petroleum in ancient texts.

Tar sand   See Bituminous sand.

Thermal coke   The carbonaceous residue formed as a result of a noncatalytic thermal process; the Conradson carbon residue; the Ramsbottom carbon residue.

Thermal cracking   A process that decomposes, rearranges, or combines hydrocarbon molecules by the application of heat, without the aid of catalysts.

Thermal process   Any refining process that utilizes heat, without the aid of a catalyst.

Topped crude   Petroleum that had volatile constituents removed up to a certain temperature, for example, 250°C + (480°F +) topped crude; not always the same as a residuum (*q.v.*).

Topping   The distillation of crude oil to remove low boiling fractions only.

Trace element   Those elements that occur at very low levels in a given system.

Trillion 1×10^{12}

True boiling point (True boiling range)   The boiling point (boiling range) of a crude oil fraction or a crude oil product under standard conditions of temperature and pressure.

Tube-and-tank cracking   An older liquid-phase thermal cracking process.

Ultimate analysis   Elemental composition.

Upgrading   The conversion of petroleum to value-added saleable products.

Vacuum distillation   Distillation (*q.v.*) under reduced pressure.

Vacuum residuum A residuum (*q.v.*) obtained by distillation of a crude oil under vacuum (reduced pressure); that portion of petroleum which boils above a selected temperature such as 510°C (950°F) or 565°C (1050°F).

Visbreaking A process for reducing the viscosity of heavy feedstocks by controlled thermal decomposition.

Viscosity A measure of the ability of a liquid to flow or a measure of its resistance to flow; the force required to move a plane surface of area 1 m^2 over another parallel plane surface 1 m away at a rate of 1 m/s when both surfaces are immersed in the fluid.

Watson characterization factor See Characterization factor.

Zeolite A crystalline aluminosilicate used as a catalyst and having a particular chemical and physical structure.